The Institution of Civil Engineers

# The New Engineering Subcontract

The form of subcontract for use
with the New Engineering Contract

 Thomas Telford, London

Published for the Institution of Civil Engineers by Thomas Telford Services Ltd, Thomas Telford House, 1 Heron Quay, London E14 4JD.

The New Engineering Contract is published as a series of documents of which this is one.

ISBN (series) 0 7277 1664 6

ISBN (this document) 0 7277 1942 4

Consultative edition 1991
First edition 1993

The New Engineering Contract has been designed and drafted by Dr Martin Barnes of Coopers and Lybrand with the assistance of Professor J. G. Perry of The University of Birmingham, T. W. Weddell of Travers Morgan Management, T. H. Nicholson, Consultant to the Institution of Civil Engineers, A. Norman of the University of Manchester Institute of Science and Technology and P. A. Baird, Corporate Contracts Consultant, Eskom, South Africa.

British Library Cataloguing in Publication Data for this publication is available from the British Library.

Printed and bound in Great Britain by Staples Printers Rochester Ltd, Rochester, Kent.

# CONTENTS

# SCHEDULE OF OPTIONS

The strategy for choosing the form of subcontract starts with a decision between five main options, one of which must be chosen.

| | |
|---|---|
| Option A | Priced subcontract with activity schedule |
| Option B | Priced subcontract with bill of quantities |
| Option C | Target subcontract with activity schedule |
| Option D | Target subcontract with bill of quantities |
| Option E | Cost reimbursable subcontract |

The following secondary options should then be considered. It is not necessary to use any of them. Any combination other than those stated may be used.

| | |
|---|---|
| Option G | Performance bond |
| Option H | Parent company guarantee |
| Option J | Advanced payment to the *Subcontractor* |
| Option K | Multiple currencies (not to be used with Options C, D and E ) |
| Option L | Sectional Completion |
| Option M | Limitation of the *Subcontractor*'s liability for design to reasonable skill and care |
| Option N | Price adjustment for inflation (not to be used with Option E ) |
| Option P | Retention |
| Option Q | Bonus for early Completion |
| Option R | Delay damages |
| Option S | Low performance damages |
| Option T | Changes in the law |
| Option U | Special conditions of subcontract |

# THE NEW ENGINEERING SUBCONTRACT

## CORE CLAUSES

## 1. General

**Actions** **10**

10.1 The *Contractor*, the *Subcontractor*, and the *Adjudicator* shall act as stated in this subcontract.

**Identified and defined** **11**
**terms** 11.1 In these conditions of subcontract, terms identified in the Subcontract Data are in italics and defined terms have capital initials.

11.2 (1) The Parties are the *Contractor* and the *Subcontractor*.

(2) Others are people or organisations who are not the *Employer*, the *Adjudicator*, the *Subcontractor*, the *Contractor* or any employee, Subsubcontractor or supplier of the *Subcontractor*.

(3) The Subcontract Date is the date when this subcontract came into existence.

(4) To Provide the Subcontract Works means to do the work necessary to complete the *subcontract works* in accordance with this subcontract and all incidental work, services and actions which this subcontract requires.

(5) Subcontract Works Information is information which

- specifies and describes the *subcontract works* or
- states how the *Subcontractor* Provides the Subcontract Works and is either

  - in the documents which the Subcontract Data states it is in or
  - in an instruction given in accordance with this subcontract.

(6) Site Information is information which

- describes the Site and its surroundings and
- is in the documents which the Subcontract Data states it is.

(7) The Site is the area within the *boundaries of the site* and the volumes above and below it which are affected by work included in this subcontract.

(8) The Working Areas are the *working areas* unless later changed in accordance with this subcontract.

(9) A Subsubcontractor is a person or corporate body who has a contract with the *Subcontractor* to provide part of the *subcontract works* or to supply Plant and Materials which he has wholly or partly designed specifically for the *subcontract works*.

(10) Plant and Materials are items to be included in the *subcontract works*.

(11) Equipment is items provided by the *Subcontractor* and used by him to

Provide the Subcontract Works and which the Subcontract Works Information does not require him to include in the *subcontract works*.

(12) The Subcontract Completion Date is the *subcontract completion date* unless later changed in accordance with this subcontract.

(13) Completion is the date, decided by the *Contractor*, when the *Subcontractor* has done all the work which the Subcontract Works Information states he is to do by the Subcontract Completion Date and has corrected notified Defects which would have prevented the *Contractor* or the *Employer* using the *subcontract works*.

(14) The Accepted Programme is the latest programme accepted by the *Contractor*. If no programme has been accepted by the *Contractor*, a programme identified in the Subcontract Data is the Accepted Programme.

(15) A Defect is

- a part of the *subcontract works* which is not completed in accordance with the Subcontract Works Information or
- a part of the *subcontract works* designed by the *Subcontractor*, which is not in accordance with

  - the applicable law or
  - the *Subcontractor*'s design which has been accepted by the *Contractor*.

(16) The Defects Certificate is either a list of Defects notified before the *defects date* which the *Subcontractor* has not corrected or, if there are no such Defects, a statement that there are none.

(17) Risk Transfer is the earlier of the date when the *Contractor* certifies that he has taken over the *subcontract works* and the date when he certifies termination.

**Interpretation  12**

12.1    In this subcontract, except where the context shows otherwise, words in the singular also mean the plural and the other way round and words in the masculine also mean the feminine and neuter.

**Communications  13**

13.1    Each instruction, certificate, submission, proposal, record, acceptance and notification which this subcontract requires is communicated in a form which can be read, copied and recorded. Writing is in the *language of this subcontract*.

13.2    A communication has effect when it is received at the last address notified by the recipient for receiving communications or, if none is notified, at the address of the recipient stated in the Subcontract Data.

13.3    A reply to a communication between the *Contractor* and the *Subcontractor* to which this subcontract requires a reply is made within the *period for reply*.

13.4    The *Contractor* replies to a communication submitted or resubmitted to him by the *Subcontractor* for acceptance. If his reply is not acceptance, he states his reasons and the *Subcontractor* resubmits the communication within the *period for reply* taking account of these reasons. A reason for witholding acceptance is that more information is needed in order to assess the *Subcontractor*'s submission fully.

13.5    The *Contractor* may extend the *period for reply* to a communication if the *Contractor* and the *Subcontractor* agree to the extension before the reply is due. The *Contractor* notifies the extension which has been agreed to the *Subcontractor*.

13.6 The *Contractor* issues his certificates to the *Subcontractor*.

13.7 Information which this subcontract requires to be notified is communicated separately from other communications.

**The *Contractor*** **14**

14.1 The *Contractor's* acceptance of a communication from the *Subcontractor* or of his work does not change the *Subcontractor's* responsibility to Provide the Subcontract Works or his liability for his design.

14.2 The *Contractor*, after notifying the *Subcontractor*, may delegate any of their actions and may cancel any delegation. A reference to an action of the *Contractor* in this subcontract includes an action by his delegate.

14.3 The *Contractor* may give an instruction to the *Subcontractor* which changes the Subcontract Works Information.

**Adding to the *working*** **15**
**areas** 15.1 The *Subcontractor* may submit a proposal for adding to the Working Areas to the *Contractor* for acceptance. Reasons for not accepting are that the proposed addition is not necessary for Providing the Subcontract Works and that the proposed area will be used for work not in this subcontract.

**Early warning** **16**

16.1 The *Subcontractor* and the *Contractor* give an early warning by notifying the other as soon as either becomes aware of any matter which could increase the total of the Prices, delay Completion or impair the performance of the *subcontract works* in use.

16.2 Either the *Contractor* or the *Subcontractor* may instruct the other to attend an early warning meeting. Each may instruct other people to attend if the other agrees.

16.3 At an early warning meeting, those who attend co-operate in

- making and considering proposals for how the effect of each matter which has been notified as an early warning can be avoided or reduced and
- deciding upon actions which they will take and who, in accordance with this subcontract, will take them.

16.4 The *Contractor* records the proposals considered and decisions taken at an early warning meeting and gives a copy of his record to the *Subcontractor*.

**Ambiguities and** **17**
**inconsistencies** 17.1 The *Contractor* or the *Subcontractor* notifies the other as soon as either becomes aware of an ambiguity or inconsistency in or between the documents which are part of this subcontract. The *Contractor* gives an instruction resolving the ambiguity or inconsistency.

**Health and safety** **18**

18.1 The *Subcontractor* acts in accordance with the health and safety requirements stated in the Subcontract Works Information.

**Illegal and impossible** **19**
**requirements** 19.1 The *Subcontractor* notifies the *Contractor* as soon as he becomes aware that the Subcontract Works Information requires him to do anything which is illegal or impossible. If the *Contractor* agrees, he gives an instruction to change the Subcontract Works Information appropriately.

# 2 The *Subcontractor*'s main responsibilities

**Providing the Subcontract Works** 20

20.1 The *Subcontractor* Provides the Subcontract Works in accordance with the Subcontract Works Information.

**The *Subcontractor*'s design** 21

21.1 The *Subcontractor* designs the parts of the *subcontract works* which the Subcontract Works Information states he is to design. His design complies with the Subcontract Works Information.

21.2 The *Subcontractor* submits the particulars of his design as the Subcontract Works Information requires to the *Contractor* for acceptance. Reasons for not accepting are that the design does not comply with the Subcontract Works Information and that it does not comply with the applicable law. The *Subcontractor* does not proceed with the relevant work until the *Contractor* has accepted his design.

21.3 The *Subcontractor* may submit his design for acceptance in parts if the design of each part can be assessed fully.

21.4 The *Subcontractor* indemnifies the *Contractor* and the *Employer* against claims, compensation and costs due to the *Subcontractor* infringing a patent or copyright.

21.5 The *Subcontractor*'s liability to the *Contractor* for his design is limited after Completion of the whole of the *subcontract works* to the amount stated in the Subcontract Data.

**Using the *Subcontractor*'s design** 22

22.1 The *Employer* and the *Contractor* may use and copy the *Subcontractor*'s design for installing, constructing, re-installing or reconstructing the *subcontract works* unless otherwise stated in the Subcontract Works Information and for other purposes as stated in the Subcontract Works Information.

**Design of Subcontract Equipment** 23

23.1 The *Subcontractor* submits particulars of the design of an item of Subcontract Equipment to the *Contractor* for acceptance if the *Contractor* instructs him to. Reasons for not accepting are that the design of the item will not allow the *Subcontractor* to Provide the Subcontract Works in accordance with

- the Subcontract Works Information,
- the *Subcontractor*'s design which the *Contractor* has accepted or
- the applicable law.

**People** 24

24.1 The *Subcontractor* either employs each key person named to do the job stated in the Subcontract Data or employs a replacement person who has been accepted by the *Contractor*. The *Subcontractor* submits the name, relevant qualifications and experience of a proposed replacement person to the *Contractor* for acceptance. A reason for not accepting the person is that his relevant qualifications and experience are not as good as those of the person who is to be replaced.

24.2 The *Contractor* may, having stated his reasons, instruct the *Subcontractor* to remove an employee. The *Subcontractor* then arranges that, after one day, the employee has no further connection with the work included in this subcontract.

**Co-operation**    **25**

25.1    The *Subcontractor* co-operates with Others in obtaining and providing information which they need in connection with the *subcontract works*. He shares the Working Areas with Others as stated in the Subcontract Works Information.

**Assignment and**    **26**
**subsubcontracting**    26.1    Neither Party assigns any benefit in a part or the whole of this subcontract. If the *Subcontractor* subsubcontracts work, he is responsible for performing this subcontract as if he had not subsubcontracted. This subcontract applies as if a Subsubcontractor's employees and equipment were the *Subcontractor*'s.

26.2    The *Subcontractor* submits the name of each proposed Subsubcontractor to the *Contractor* for acceptance. A reason for not accepting the Subsubcontractor is that his appointment will not allow the *Subcontractor* to Provide the Subcontract Works. The *Subcontractor* does not appoint a proposed Subsubcontractor until the *Project Manager* has accepted him.

**Approval from Others**    **27**

27.1    The *Subcontractor* obtains approval of his design from others where necessary.

**Access to the work**    **28**

28.1    The *Subcontractor* provides access for the *Contractor*, and others notified by the *Contractor* to work being done for this subcontract and to stored Plant and Materials.

**Instructions**    **29**

29.1    The *Subcontractor* obeys an instruction which the *Contractor* gives him and which is in accordance with this Subcontract.

# 3 Time

**Starting and Completion** 30

30.1 The *Subcontractor* does not start work on the Site until the first *subcontract possession date* and does the work so that Completion is on or before the Subcontract Completion Date.

30.2 The *Contractor* certifies Completion within two weeks of Completion.

**The Accepted** 31
**Programme** 31.1 If the Accepted Programme is not identified in the Subcontract Data, the *Subcontractor* prepares and submits a programme to the *Contractor* for acceptance within the period stated in the Subcontract Data.

31.2 Reasons for not accepting a programme (including a revised programme) are that the *Subcontractor*'s plans which it shows are not practicable and that the programme does not

- include the information which this subcontract requires
- represent the *Subontractor*'s plans realistically or
- show realistic provisions for

  - float and other risk allowances
  - health and safety requirements
  - other requirements of the Subcontract Works Information or
  - the procedures set out in this subcontract.

31.3 The Accepted Programme includes

- the *subcontract starting date*, *subcontract possession dates* and Subcontract Completion Date
- for each operation a method statement which identifies the Equipment and other resources which the *Subcontractor* plans to use
- planned Completion
- the order and timing of

  - the operations which the *Subcontractor* plans to do in order to Provide the Subcontract Works and
  - the work of others either as stated in the Subcontract Works Information or as later agreed with them by the *Subcontractor*

- the dates when the *Subcontractor* plans to complete work needed to allow Others to do their work
- the dates when, in order to Provide the Subcontract Works in accordance with his programme, the *Subcontractor* will need

  - possession of a part of the Site if later than its *subcontract possession date*
  - acceptances and
  - Plant and Materials and other things to be provided by the *Employer* and the *Contractor* and

- other information which the Subcontract Works Information requires the *Subcontractor* to show on the Accepted Programme.

**Revising the programme** 32

32.1 Each revised programme shows

- the actual progress achieved on each operation since the last revision and its effect upon the timing of the remaining work

- the effects of implemented compensation events and of notified early warning matters
- how the *Subcontractor* plans to deal with any delays and to correct notified Defects and
- any other changes which the *Subcontractor* proposes to make to the Accepted Programme.

32.2 The *Subcontractor* submits a revised programme to the *Contractor* for acceptance

- within the *period for reply* after the *Contractor* has instructed him to
- when the *Subcontractor* chooses to and, in any case,
- at no longer interval than the interval stated in the Subcontract Data from the *subcontract starting date* until Completion of the whole of the *subcontract works*.

**Possession of the Site** 33

33.1 The *Contractor* gives possession of each part of the Site to the *Subcontractor* on or before the later of its *subcontract possession date* and the date for possession shown on the Accepted Programme.

33.2 While the *Subcontractor* has possession of a part of the Site, the *Contractor* gives the *Subcontractor* access to and use of it and the *Contractor* and the *Subcontractor* provide facilities and services as stated in the Subcontract Works Information. Any cost incurred by the *Contractor* as a result of the *Subcontractor* not providing the facilities and services stated is assessed by the *Contractor* and paid by the *Subcontractor*.

**Instructions to stop or** 34
**not to start work** 34.1 The *Contractor* may instruct the *Subcontractor* to stop or not to start any work and may later instruct him that he may re-start or start it.

**Taking over** 35

35.1 Possession of each part of the Site returns to the *Contractor* when he takes over the part of the *subcontract works* which occupies it. Possession of the whole Site returns to the *Contractor* when the *Contractor* certifies termination.

35.2 The *Contractor* need not take over the *subcontract works* before the Subcontract Completion Date if it is stated in the Subcontract Data that he is not willing to do so. Otherwise the *Contractor* takes over the *subcontract works* not more than two weeks after Completion.

35.3 The *Employer* may use any part of the *subcontract works* before Completion has been certified. If he does so, the *Contractor* takes over the part of the *subcontract works* before the *Employer* begins to use it.

35.4 The *Contractor* certifies the date upon which he takes over any part of the *subcontract works* within two weeks of the date.

**Acceleration** 36

36.1 The *Contractor* may instruct the *Subcontractor* to submit a quotation for an acceleration to achieve Completion before the Subcontract Completion Date. A quotation for an acceleration comprises proposed changes to the Prices and the Subcontract Completion Date and a revised programme.

36.2 The *Subcontractor* submits a quotation or gives his reasons for not doing so within the *period for reply*.

# 4 Testing and Defects

**Tests and inspections** **40**

40.1 This clause only governs tests and inspections required by the Subcontract Works Information and the applicable law.

40.2 The *Subcontractor*, the *Contractor* and the *Employer* provide materials, facilities and samples for tests and inspections as stated in the Subcontract Works Information.

40.3 The *Subcontractor* and the *Contractor* notify the other of each of their tests and inspections before it starts and afterwards notify its results. The *Subcontractor* notifies the *Contractor* in time for a test or inspection to be arranged and done before doing work which would obstruct the test or inspection. The *Contractor* and the *Supervisor* may watch any test done by the *Subcontractor*.

40.4 If a test or inspection shows that any work has a Defect, the *Subcontractor* corrects the Defect and the test or inspection is repeated.

40.5 The *Contractor* does his tests and inspections without causing unnecessary delay to the work or to a payment which is conditional upon a test or inspection being successful. A payment which is conditional upon a *Contractor*'s or *Supervisor*'s test or inspection being successful becomes due at the later of the *defects date* and the end of the last *defect correction period* if

- the *Contractor* or the *Supervisor* has not done the test or inspection and
- the delay to the test or inspection is not the *Subcontractor*'s fault.

40.6 The *Contractor* assesses the cost incurred by him in repeating a test or inspection after a Defect is found. The *Subcontractor* pays the amount assessed.

**Testing and inspection** **41**
**before delivery** 41.1 The *Subcontractor* may not bring Plant and Materials to the Working Areas which the Subcontract Works Information states are to be tested or inspected before delivery until the *Contractor* has notified the *Subcontractor* that they have passed the test or inspection.

**Searching and notifying** **42**
**Defects** 42.1 The *Contractor* may instruct the *Subcontractor* to search. He gives his reason for the search with his instruction. Searching may include

- uncovering, dismantling, re-covering and re-erecting work,
- providing facilities, materials and samples for tests and inspections done by the *Contractor* or the *Supervisor* and
- doing tests and inspections which the Subcontract Works Information does not require.

42.2 The *Subcontractor* notifies the *Contractor* of each Defect which he finds. The *Contractor* notifies the *Subcontractor* of each Defect which he finds. No Defects are notified after the *defects date*.

**Correcting Defects** **43**

43.1 The *Subcontractor* corrects Defects whether or not the *Contractor* notifies him of them. The *Subcontractor* corrects notified Defects before the end of the *defect correction period*. This period begins at Completion for Defects notified before Completion and when the Defect is notified for other Defects.

43.2 The *Contractor* issues the Defects Certificate at the later of the *defects date* and the end of the last *defect correction period*.

43.3     The *Contractor* makes arrangements to give the *Subcontractor* the access to and use of those parts of the Site which the *Subcontractor* needs to correct a Defect after the *Contractor* has taken over any part of the *subcontract works*. If the *Contractor* has not arranged suitable access and use within the *Defect correction period*, he extends the period for the Defect as necessary.

**Accepting Defects**    **44**

44.1     The *Subcontractor* or the *Contractor* may propose to the other that the Subcontract Works Information should be changed so that a Defect does not have to be corrected.

44.2     If the *Subcontractor* and the *Contractor* are prepared to consider the change, the *Subcontractor* submits a quotation for reduced Prices or an earlier Subcontract Completion Date or both to the *Contractor* for acceptance. If the *Contractor* accepts the quotation, he gives an instruction to change the Subcontract Works Information, the Prices and the Subcontract Completion Date accordingly.

**Uncorrected Defects**    **45**

45.1     If the *Subcontractor* has not corrected a Defect within its *defect correction period*, the *Contractor* assesses the cost of having the Defect corrected by other people and the *Subcontractor* pays this amount.

# 5 Payment

**Assessing the amount** **50**
**due** 50.1 The *Contractor* assesses the amount due at each assessment date. The first assessment date is decided by the *Contractor* to suit the procedures of the Parties and is not later than the *assessment interval* after the *subcontract starting date*. Later assessment dates occur

- at the end of each *assessment interval* until Completion of the whole of the *subcontract works*,
- at Completion of the whole of the *subcontract works*,
- when the *Contractor* issues the Defects Certificate and
- after Completion of the whole of the *subcontract works,*

    - when an amount due is corrected and
    - when a payment is made late.

50.2 The amount due is the Price for Work Done to Date plus other amounts to be paid to the *Subcontractor* less amounts to be paid by or retained from the *Subcontractor*. Any value added tax or sales tax which the law requires the *Contractor* to pay to the *Subcontractor* is included in the amount due.

50.3 If there is no Accepted Programme and the *Subcontractor* has not submitted a programme to the *Contractor* for acceptance which contains the information which this subcontract requires, half the Price for Work Done to Date is retained from the *Subcontractor* in assessments of the amount due.

50.4 In assessing the amount due, the *Contractor* considers any application for payment which the *Subcontractor* has submitted on or before the assessment date. The *Contractor* gives the *Subcontractor* details of how the amount due has been made up.

50.5 The *Contractor* corrects any wrongly assessed amount due in a later payment certificate.

**Payment** **51**
51.1 The *Contractor* certifies a payment within two weeks of each assessment date. The first payment is the amount due, other payments are the change in the amount due since the last payment certificate. A payment is made by the *Subcontractor* to the *Contractor* if the change reduces the amount due. Other payments are made by the *Contractor* to the *Subcontractor*. Payments are in the *currency of this subcontract* unless otherwise stated in this subcontract.

51.2 Each certified payment is made within six weeks of the assessment date or, if a different period is stated in the Subcontract Data, within the period stated. If a payment is late, interest is paid on the late payment. Interest is assessed from the date by which the late payment should have been made until the date when the late payment is made and is included in the first assessment after the late payment is made.

51.3 If an amount due is corrected in a later certificate either by the *Contractor* in relation to a compensation event or following a decision of the *Adjudicator* or an arbitrator, interest on the correcting amount is paid. Interest is assessed from the date that the incorrect amount was certified until the date that the correcting amount is certified and is included in the assessment which includes the correcting amount.

51.4 If the *Contractor* does not issue a certificate which he should issue, interest is paid on the amount which he should have certified. Interest is assessed from the date by which he should have certified the amount until the date when he certifies the amount and it is included in the amount then certified.

51.5 Interest is calculated at the *interest rate* and is compounded over the period for which interest is assessed.

**Actual Cost 52**

52.1 All the *Subcontractor*'s costs which are not included in Actual Cost are deemed to be included in the *fee percentage*. Payments by the *Subcontractor* included in Actual Cost are at open market or competitively tendered prices with all discounts, rebates and taxes which can be recovered deducted.

# 6 Compensation events

**Compensation events 60**

60.1     The following are compensation events.

(1) The *Contractor* gives an instruction changing the Subcontract Works Information except

- a change made in order to accept a Defect or
- a change to the Subcontract Works Information provided by the *Subcontractor* for his design which is made at his request or to comply with Subcontract Works Information provided by the *Contractor*.

(2) The *Contractor* does not give possession of a part of the Site by the later of its *subcontract possession date* and the date shown on the Accepted Programme.

(3) The *Contractor* does not provide something which he is to provide by the date for providing it shown on the Accepted Programme.

(4) The *Contractor* gives an instruction to stop or not to start any work.

(5) Others do not work within the times shown on the Accepted Programme or do not work within the conditions stated in the Subcontract Works Information.

(6) The *Contractor* does not reply to a communication from the *Subcontractor* within the period required by this contract.

(7) The *Contractor* gives an instruction for dealing with an object of value or of historical or other interest found within the Site.

(8) The *Contractor* changes a decision which he has previously communicated to the *Subcontractor*.

(9) The *Contractor* withholds an acceptance (other than an acceptance of a quotation for acceleration or for accepting a Defect) for a reason not stated in this subcontract.

(10) The *Contractor* instructs the *Subcontractor* to search and no Defect is found except when the *Subcontractor* did not notify the *Contractor* in time for a required test or inspection to be arranged and done before doing work which obstructed the test or inspection.

(11) A test or inspection done by the *Contractor* or the *Supervisor* causes unnecessary delay.

(12) The *Subcontractor* encounters physical conditions within the Site, other than weather conditions, which, at the Subcontract Date, an experienced subcontractor would have judged to have such a small chance of occurring that it would have been unreasonable for him to have allowed for them.

(13) Weather is recorded within a calendar month and before the Subcontract Completion Date for the whole of the *subcontract works* at the place stated in the Subcontract Data which one of the *weather measurements*, when compared with the *weather data,* shows has occurred on average less frequently than once in ten years.

(14) An *Employer*'s risk event or a *Contractor*'s risk event retained by the *Contractor* occurs.

(15) The *Contractor* uses part of the *subcontract works* before both Completion and the Subcontract Completion Date.

(16) The *Contractor* does not provide materials, facilities and samples for tests as stated in the Subcontract Works Information.

60.2  In judging the expected physical conditions, the *Subcontractor* is assumed to have taken into account

- the Site Information,
- publicly available information referred to in the Site Information,
- information obtainable from a visual inspection of the Site and
- other information which an experienced subcontractor could reasonably be expected to have or to obtain.

60.3  If there is an inconsistency within the Site Information (including the information referred to in it), the *Subcontractor* is assumed to have taken into account the physical conditions more favourable to doing the work.

**Notifying compensation events**  **61**

61.1  For compensation events which arise from the *Contractor* giving an instruction or changing an earlier decision, the *Contractor* notifies the *Subcontractor* of the compensation event at the time of the event. He also instructs the *Subcontractor* to submit quotations unless the event arises from a fault of the *Subcontractor* or quotations have already been submitted. The *Subcontractor* puts the instruction or changed decision into effect.

61.2  The *Contractor* may instruct the *Subcontractor* to submit quotations for a proposed instruction or changed decision. The *Subcontractor* does not put a proposed instruction or changed decision into effect.

61.3  For compensation events not notified by the *Contractor*, the *Subcontractor* notifies the *Contractor* why he considers that the compensation event has happened or is expected to happen. The *Subcontractor* may not notify a compensation event more than one week after he became aware of it.

61.4  For compensation events notified by the *Subcontractor*, the *Contractor* decides

- whether the compensation event

  - arises from a fault of the *Subcontractor*
  - has happened or should be expected to happen
  - could have an effect upon Actual Cost or upon Completion and

- whether the *Subcontractor* gave an early warning of the compensation event which an experienced subcontractor could have given.

61.5  The *Contractor* notifies the *Subcontractor* of his decisions within five weeks of the *Subcontractor's* notification. If he decides that the compensation event did not arise from a fault of the *Subcontractor*, that it has happened or should be expected to happen and that it could have an effect upon Actual Cost or Completion, he instructs the *Subcontractor* to submit quotations at the time he notifies his decisions. If he decides otherwise, the notification of the compensation event is deemed to have been withdrawn.

61.6  If the *Contractor* decides that the nature of a compensation event is too uncertain for its effect to be forecast reasonably, he states assumptions about its nature on which assessment is to be based when he instructs the *Subcontractor* to submit quotations. If any of these assumptions is later found to have been wrong, the *Contractor* notifies a correction as a compensation event.

61.7  A compensation event may not be notified after the *defects date*.

**Quotations for compensation events**  **62**

62.1  The *Contractor* may instruct the *Subcontractor* to submit alternative quotations based upon different ways of dealing with the compensation event which are practicable. The *Subcontractor* submits the required quotations to the *Contractor* and may submit quotations for other methods of dealing with the compensation event which he considers practicable.

62.2 Quotations for compensation events comprise proposed changes to the Prices and any delay to the Subcontract Completion Date assessed by the *Subcontractor*. He submits details of his assessment with each quotation. If the programme for remaining work is affected by the compensation event, the *Subcontractor* includes a revised programme showing the effect in his quotation.

62.3 The *Subcontractor* submits quotations within one week of being instructed to do so by the *Contractor*. The *Contractor* replies within five weeks of the submission. His reply is

- an instruction to submit a revised quotation,
- an acceptance of a quotation,
- a notification that a proposed instruction or changed decision will not be notified or
- a notification that he will be making his own assessment.

62.4 The *Contractor* instructs the *Subcontractor* to submit a revised quotation only after explaining his reasons for doing so to the *Subcontractor*. The *Subcontractor* submits the revised quotation within one week of being instructed to do so.

**Assessing compensation events**

**63**

63.1 The changes to the Prices are assessed as the effect of the compensation event upon

- the Actual Cost of the work already done,
- the forecast Actual Cost of the work not yet done and
- the resulting Fee.

63.2 If the effect of a compensation event is to reduce the total Actual Cost, the Prices are not reduced except as stated in this subcontract. If the effect of a compensation event which is a change to the Subcontract Works Information or a correction of an assumption stated by the *Contractor* for assessing an earlier compensation event is to reduce the total Actual Cost, the Prices are reduced.

63.3 A delay to the Subcontract Completion Date is assessed as the length of time that, due to the compensation event, planned Completion is later than planned Completion as shown on the Accepted Programme.

63.4 If the *Contractor* has notified the *Subcontractor* of his decision that the *Subcontractor* did not give an early warning of a compensation event which an experienced subcontractor could have given, the compensation event is assessed taking into account any savings of Actual Cost and time which would have occurred if the *Subcontractor* had given early warning.

63.5 Allowances for cost-increasing and delaying factors which have a significant chance of occurring and which are at the *Subcontractor*'s risk under this subcontract are included in forecasts of Actual Cost and Completion.

63.6 Assessments are based upon the assumptions that the *Subcontractor* reacts competently and promptly to the compensation event, that the additional Actual Cost and time due to the event are reasonably incurred and that the Accepted Programme can be changed.

63.7 A compensation event which is an instruction to change the Subcontract Works Information in order to resolve an ambiguity or inconsistency is assessed as follows. If Subcontract Works Information provided by the *Contractor* is changed, the effect of the compensation event is assessed as if the Prices and the Subcontract Completion Date were for the interpretation most favourable to the *Subcontractor*. If Subcontract Works Information provided by the *Subcontractor* is changed, the effect of the compensation event is assessed as if the Prices and the Subcontract Completion Date were for the interpretation most favourable to the *Contractor*.

**The *Contractor*'s assessments**    **64**

64.1    The *Contractor* may assess a compensation event

- if the *Subcontractor* has not submitted a required quotation and details of his assessment within the time allowed,
- if the *Contractor* decides that the *Subcontractor* has not assessed the compensation event correctly in a quotation which is otherwise acceptable,
- if, at the time of notifying the compensation event, the *Subcontractor* has not submitted a programme which this subcontract requires him to submit or
- if the reason for the *Contractor* not accepting the latest programme submitted by the *Subcontractor* is one of the reasons stated in this subcontract.

64.2    If there is no Accepted Programme or if the Accepted Programme is out of date, the *Contractor* assesses the compensation event using his own assessment of the programme for the remaining work.

64.3    The *Contractor* notifies the *Subcontractor* of his assessment of a compensation event and gives him details of it within three weeks of the need for his assessment becoming apparent.

**Implementing compensation events**    **65**

65.1    The *Contractor* implements each compensation event by notifying the *Subcontractor* of the quotation which he has accepted or of his own assessment. He implements the compensation event when he accepts a quotation or completes his own assessment or when the compensation event occurs, whichever is latest.

65.2    The assessment of a compensation event is not reviewed in the light of information about its actual effect upon Actual Cost or upon the timing of the work which becomes available only after the *Subcontractor* has or should have submitted his quotations.

# 7 Title

**The *Contractor*'s title to**  **70**
**Equipment, Plant and**  70.1   Whatever title the *Subcontractor* has to Equipment or Plant and Materials
**Materials**     which is outside the Working Areas passes to the *Contractor* if the *Contractor* has made a payment for the Equipment or Plant and Materials or the *Subcontractor* has marked it as for this subcontract.

    70.2   Whatever title the *Subcontractor* has to Equipment or Plant and Materials passes to the *Contractor* if it has been brought within the Working Areas and passes back to the *Contractor* if it is removed from the Working Areas with the permission of the *Contractor*.

**Plant and Materials**  **71**
**outside the Working**  71.1   The *Contractor* marks Plant and Materials which are outside the Working
**Areas**     Areas if this subcontract provides for payment for them and the *Subcontractor* has shown him that any requirements of the Subcontract Works Information with respect to marking have been satisfied.

**Removing Equipment**  **72**
    72.1   The *Subcontractor* removes Equipment from the Site when it is no longer needed unless the *Contractor* allows it to be left in the *subcontract works*.

**Objects and materials**  **73**
**found within the Site**  73.1   The *Subcontractor* has no title to an object of value or of historical or other interest found within the Site. The *Subcontractor* notifies the *Contractor* when such an object is found and the *Contractor* instructs the *Subcontractor* how to deal with it. The *Subcontractor* may not move the object without instructions.

    73.2   The *Subcontractor* has title to materials from excavation and demolition removed from the Site only as stated in the Subcontract Works Information.

# 8 Risks and insurance

**The *Employer*'s and** **80**
**The *Contractor*'s risks** 80.1 The *Employer* in the main contract carries the risks which are stated in this subcontract to be *Employer*'s risks. The *Contractor* carries the risks which are stated to be *Contractor*'s risks, except those risks which are transferred to the *Subcontractor* under this subcontract.

80.2 From the main contract starting date until the main contract defects certificate has been issued, the following are *Employer*'s risks.

- The risk of claims, proceedings, compensation and costs payable for personal injury, death or loss of or damage to property (excluding the main contract *works*, plant and materials) which are due to

  - use or occupation of the Site by the main contract *works* or for the purpose of the main contract *works* which is the unavoidable result of the main contract *works* or
  - negligence, breach of statutory duty or interference with any legal right by the *Employer* or by any person employed by or subcontracted to him except the *Contractor*.

- The risk of damage to the main contract *works*, plant and materials to the extent that it is due to a fault of the *Employer* or in his design.
- The risks of war and of radioactive contamination.
- Other *Employer*'s risks stated in the contract data of the main contract.

80.3 From main contract risk transfer until the main contract defects certificate has been issued, the risk of loss of or damage to the main contract *works*, plant and materials is an *Employer*'s risk except loss or damage due to

- a main contract defect which existed at main contract risk transfer,
- an event occurring before main contract risk transfer which was not itself an *Employer*'s risk or
- the activities of the *Contractor* on the Site after main contract risk transfer.

**The *Contractor*'s and** **81**
**Subcontractor's risks** 81.1 From the main contract starting date until the main contract defects certificate has been issued, the risks of personal injury, death and loss of or damage to property which are not *Employer*'s risks are *Contractor*'s risks.

81.2 Under this subcontract, the *Contractor*'s risks in the main contract, except where they are retained by the *Contractor* as stated in the Subcontract Data, are transferred to and become the *Subcontractor*'s risks in respect of the *subcontract works* and for the periods relating to the *subcontract starting date*, Risk Transfer and Defects Certificate.

**Repairs** **82**
82.1 Before the Defects Certificate has been issued, the *Contractor* may instruct the *Subcontractor* to replace loss of, or repair damage to, the *subcontract works*, Plant and Materials due to an *Employer*'s or *Contractor*'s risk event.

82.2 The *Subcontractor* promptly replaces loss of and repairs damage to the *subcontract works*, Plant and Materials due to a *Subcontractor*'s risk event.

**Liability for personal** **83**
**injury, death and loss of** 83.1 Each Party indemnifies the other against claims, compensation and costs for
**or damage to property** personal injury, death and loss of or damage to property (excluding the *subcontract works*, Plant and Materials) due to an event which is at his risk.

19

The *Contractor* indemnifies the *Subcontractor* against all claims and liabilities against which the *Employer* idemnifies the *Contractor* under the main contract.

83.2 The liability of each Party to indemnify the other is reduced in proportion to the extent that events which were at the other Party's risk contributed to the loss, damage, personal injury or death and taking into account each Party's responsibilities under this subcontract.

**Insurance cover 84**

84.1 The *Subcontractor* provides the following insurances and those stated in the Subcontract Data unless it is stated in the Subcontract Data that the *Employer* or *Contractor* provides them. The insurances are in the joint names of the Parties and provide cover from the *subcontract starting date* until the Defects Certificate has been issued. For events which are due to the *Subcontractor*'s risks. The cover stated is the minimum for any one event.

1. Event: loss of or damage to the *subcontract works*, Plant and Materials.
   Cover: 115% of the replacement cost on a value indemnity basis with the maximum deductible stated in the Subcontract Data and with the amount stated in the Subcontract Data in respect of the *Subcontractor*'s faulty design.

2. Event: loss of or damage to Equipment.
   Cover: the replacement cost on a value indemnity basis with the maximum deductible stated in the Subcontract Data.

3. Event: loss of or damage to property (except the *subcontract works*, Plant and Materials and Equipment) in connection with this subcontract.
   Cover: the amount stated in the Subcontract Data with cross liability so that the insurance applies to the Parties as separate insured and with the amount stated in the Subcontract Data in respect of the *Subcontractor*'s faulty design.

4. Event: personal injury or death.
   Cover: the amount stated in the Subcontract Data with cross liability so that the insurance applies to the Parties as separate insured.

84.2 Insurance policies may exclude cover for

- the cost of correcting a Defect
- consequential loss
- payment of delay or low performance damages
- wear and tear, shortages and pilferage and
- risks relating to Equipment which the law requires to be insured.

**Insurance policies 85**

85.1 The *Subcontractor* submits policies and certificates for insurance to the *Contractor* for acceptance before the *subcontract starting date* and afterwards as the *Contractor* instructs. A reason for not accepting the policies and certificates is that they do not comply with this subcontract.

85.2 Insurance policies include a waiver by the insurers of their subrogation rights against directors and other employees of every insured except where there is fraud.

85.3 The Parties comply with the terms and conditions of the insurance policies.

85.4 Any amount not recovered from an insurer is borne by the *Employer* for his risks, by the *Contractor* for his risks or by the *Subcontractor* for his.

**If the *Subcontractor* does** **86**
**not insure** 86.1     The *Contractor* may insure a risk which this subcontract requires the *Subcontractor* to insure if the *Subcontractor* does not submit a required policy or certificate. The cost of this insurance to the *Contractor* is paid by the *Subcontractor*.

**Insurance by the** **87**
***Employer* and the** 87.1     The *Employer* provides the insurances under the main contract stated in the
***Contractor***        Subcontract Data. The *Contractor* provides the insurances stated in the Subcontract Data.

       87.2     The *Contractor* submits policies and certificates for insurances provided by the *Employer* and the *Contractor* to the *Subcontractor* for acceptance before the *subcontract starting date* and afterwards as the *Subcontractor* instructs. The *Subcontractor* accepts the policies and certificates if they comply with this subcontract.

# 9 Disputes and termination

| | | |
|---|---|---|
| **Disputes about an action taken by the *Contractor*** | **90** | |
| | 90.1 | If the *Subcontractor* believes that an action of the *Contractor* was not in accordance with this subcontract or was outside the authority given by this subcontract, he may notify the *Adjudicator* and the *Contractor* of the disputed action within four weeks of the action. Unless this subcontract has been terminated, the Parties implement the disputed action. |
| | 90.2 | Within two weeks of the notification, the *Contractor* provides the *Adjudicator* and the *Subcontractor* with the information upon which the disputed action was based. Within two weeks of receiving this information, the *Subcontractor* may provide the *Adjudicator* and the *Contractor* with any other information upon which he believes the *Contractor* should have based the disputed action. |
| | 90.3 | The *Adjudicator* decides whether the disputed action was in accordance with this subcontract and whether it was within the authority given by this subcontract. If he decides that it was not, he decides what action should have been taken and assesses any additional cost and delay which the dispute itself has or will cause to the *Subcontractor*. The *Adjudicator* makes his assessment in the same way as a compensation event is assessed. |
| **Disputes about an action not taken by the *Contractor*** | **91** | |
| | 91.1 | If the *Subcontractor* believes that the *Contractor* has not taken an action which this subcontract requires, he may notify the *Contractor*. |
| | 91.2 | If the action has not been taken within four weeks of this notification, the *Subcontractor* may notify the *Adjudicator* and the *Contractor* within a further four weeks. The *Subcontractor* may include in this notification information which he believes shows that the *Contractor* should have taken the action. Within two weeks of the notification to the *Adjudicator*, the *Contractor* may supply the *Adjudicator* with information which he believes shows that he should not have taken the action. |
| | 91.3 | The *Adjudicator* decides whether, in accordance with this subcontract, the action should or should not have been taken. If he decides that it should have been taken, the action is implemented and he assesses any additional cost and delay which the dispute itself has caused or will cause to the *Subcontractor*. The *Adjudicator* makes his assessment in the same way as a compensation event is assessed. |
| **The *Adjudicator*** | **92** | |
| | 92.1 | The *Adjudicator* notifies the *Contractor* and the *Subcontractor* of his decision, of the reason for his decision and of any assessment within four weeks of receiving the information or within a longer period which has been agreed by the *Contractor* and the *Subcontractor*. The *Contractor* implements the *Adjudicator*'s assessment as if it had resulted from a compensation event. |
| | 92.2 | If the *Adjudicator* resigns or is unable to act, the Parties choose a new adjudicator jointly. If the Parties have not chosen a new adjudicator jointly within four weeks of the *Adjudicator* resigning or becoming unable to act, any Party may ask the person stated in the Subcontract Data to choose a new adjudicator and the Parties accept his choice. The new adjudicator is appointed on terms which provide that payments to him are shared equally between the Parties and that he is not liable to the Parties for breach of duty. |
| | 92.3 | The *Subcontractor* may notify a dispute arising from a subsubcontract which is also a dispute under this subcontract to the *Adjudicator*. The *Adjudicator* then decides the two disputes as one, the Subsubcontractor becomes a party to the dispute and references to the Parties in clauses of *the conditions of* |

*subcontract* concerned with the dispute are interpreted as including the sub-subcontract.

92.4 If any dispute arising under the main contract concerns the *subcontract works*, and is referred to the adjudicator in the main contract, the *Contractor* may, by notification to the *Subcontractor*

- require him to provide information the *Contractor* may request
- require that any dispute under this subcontract on the same matter be dealt with jointly with the dispute under the main contract.

The adjudicator in the main contract then decides the two disputes as one, and the *Subcontractor* becomes a party to the dispute.

**Arbitration 93**

93.1 If the *Adjudicator* does not notify his decision within the time provided by this subcontract or if a Party disagrees with his decision, any Party to the dispute may give to the other parties a notice to refer the dispute to an arbitrator. A notice to refer a dispute to an arbitrator may not be given more than four weeks after the *Adjudicator* has or should have notified his decision.

93.2 A dispute referred to an arbitrator is conducted using the *arbitration procedure*.

**Termination 94**

94.1 If either Party wishes to terminate, he notifies the other party giving details of his reason for terminating. The *Contractor* issues a termination certificate promptly if the reason complies with this subcontract.

94.2 The *Subcontractor* may terminate only for a reason identified in the table. The *Contractor* may terminate for any reason. The procedures followed and the amounts due on termination which depend upon which Party terminates and upon his reason for terminating are identified in the table.

| Terminating Party | Reason | Procedure | Amount Due |
|---|---|---|---|
| The *Contractor* | A reason other than R1 - R18 | P1 | A1 and A3 |
| | R1 - R12, R16 | P1 and P2 | A2 |
| | R14, R15, R18 | P2 | A1 and A4 |
| The *Subcontractor* | R1 - R7, R13, R17 | P3 | A1 and A3 |
| | R14, R15, R18 | P2 | A1 and A4 |

94.3 The procedures for termination are implemented immediately after the *Contractor* has issued a termination certificate.

94.4 As soon as possible after termination, the *Contractor* certifies a final payment to or from the *Subcontractor* which is the *Contractor*'s assessment of the amount due on termination less the total of previous payments.

94.5 After a termination certificate has been issued, the *Subcontractor* does no further work necessary to complete the *subcontract works*.

**Reasons for termination 95**

95.1 Either Party may terminate if the other Party has done one of the following or its equivalent

- become bankrupt or insolvent (R1),

- had a bankruptcy order made against him (R2),
- presented his petition in bankruptcy (R3),
- made an arrangement with or an assignment in favour of his creditors (R4),
- agreed to perform this subcontract under a committee of inspection of his creditors (R5)
- or gone into liquidation other than voluntarily in order to amalgamate or reconstruct (R6) or
- assigned this subcontract (R7).

95.2 The *Contractor* may terminate if he has notified that the *Subcontractor* has defaulted in one of the following ways and not put the default right within three weeks of the notification.

- Substantially failed to comply with his obligations (R8).
- Not provided a bond or guarantee which this subcontract requires (R9).
- Appointed a Subsubcontractor for substantial work before the *Contractor* has accepted the Subsubcontractor (R10).

95.3 The *Contractor* may terminate if he has notified that the *Subcontractor* has defaulted in one of the following ways and not stopped defaulting within three weeks of the notification.

- Substantially hindered the *Employer*, the *Contractor* or Others (R11).
- Substantially broken a health or safety regulation (R12).

95.4 The *Subcontractor* may terminate if the *Contractor* has not paid an amount he has certified within thirteen weeks of the date of the certificate (R13).

95.5 Either Party may terminate if

- war or radioactive contamination has substantially affected the *Subcontractor*'s work for 26 weeks (R14) or
- the Parties have been released under the law from further performance of the whole of this subcontract (R15).

95.6 If the *Contractor* has instructed the *Subcontractor* to stop or not to start any substantial work or all work and an instruction allowing the work to restart or start has not been given within thirteen weeks,

- the *Contractor* may terminate if the instruction was due to a default by the *Subcontractor* (R16)
- the *Subcontractor* may terminate if the instruction was due to a default by the *Contractor* (R17) and
- either Party may terminate if the instruction was due to any other reason (R18).

**Procedures on termination** 96

96.1 The *Contractor* may complete the *subcontract works* himself or employ other people to do so and may use any Plant and Materials to which he has title.

96.2 The procedure on termination also includes one or more of the following as set out in the table.

| | |
|---|---|
| P1 | The *Contractor* may instruct the *Subcontractor* to leave the Site, remove any Equipment, Plant and Materials from the Site and assign the benefit of any subsubcontract or other contract related to performance of this subcontract to the *Contractor*. |
| P2 | The *Contractor* may use any Equipment to which he has title. |
| P3 | The *Subcontractor* leaves the Working Areas and removes the Equipment. |

**Payment on termination** **97**

97.1 The amount due on termination includes

- an amount due assessed as for normal payments,
- the Actual Cost for Plant and Materials

  - within the Working Areas or
  - to which the *Contractor* has title and of which the *Subcontractor* has to accept delivery,

- other Actual Cost reasonably incurred in expectation of completing the *subcontract works*,
- any amounts retained by the *Contractor* and
- a deduction of any unrepaid balance of an advanced payment.

97.2 The amount due on termination also includes one or more of the following as set out in the table.

| | |
|---|---|
| A1 | The forecast Actual Cost of removing the Equipment. |
| A2 | A deduction of the forecast of the additional cost to the *Contractor* of completing the *subcontract works*. |
| A3 | The *fee percentage* applied to<br><br>• for Options A, B, C and D, any excess of the total of the Prices at the Subcontract Date over the Price for Work Done to Date or<br>• for Option E any excess of the first forecast of the Actual Cost for the *subcontract works* over the Price for Work Done to Date less the Fee. |
| A4 | Half of A3. |

# MAIN OPTION CLAUSES

## Option A: Priced subcontract with activity schedule

**Identified and defined terms**    **11**

11.2    (18) The Fee is the amount calculated by applying the *fee percentage* to the amount of Actual Cost.

(20) The Prices are the lump sum prices for each of the activities in the *activity schedule* unless later changed in accordance with this subcontract.

(24) The Price for Work Done to Date is the total of the Prices for each group of activities which the *Subcontractor* has completed and for each activity which is not in a group which he has completed.

(28) Actual Cost is the cost of the components in the Schedule of Cost Components whether work is subsubcontracted or not.

**The Accepted Programme**    **31**

31.4    The start and finish of each activity on the *activity schedule* are shown on the Accepted Programme.

**Acceleration**    **36**

36.3    When the *Contractor* accepts a quotation for an acceleration, he changes the Subcontract Completion Date and the Prices accordingly and accepts the revised programme.

**The *activity schedule***    **54**

54.1    Information in the *activity schedule* is not Subcontract Works Information or Site Information.

54.2    If the *Subcontractor* changes a planned method of working at his discretion so that the *activity schedule* does not comply with the Accepted Programme, the *Subcontractor* submits a revision of the *activity schedule* to the *Contractor* for acceptance.

54.3    Reasons for not accepting a revision of the *activity schedule* are that

- it does not comply with the Accepted Programme,
- any changed Prices are not reasonably distributed between the activities and that
- the total of the Prices is changed.

**Assessing compensation events**    **63**

63.8    Assessments for changed Prices for compensation events are in the form of changes to the *activity schedule*.

63.10    Assessment of a compensation event for work which includes subsubcontracted work does not include fees paid or to be paid by the *Subcontractor* to Subsubcontractors.

63.11    The *Contractor* may instruct the *Subcontractor* to assess the effect of a compensation event upon Actual Cost using the short method set out in the Schedule of Cost Components if the *Subcontractor* agrees and may make his own assessments using the short method.

**Implementing** **65**
**compensation events** 65.4    The *Contractor* includes the changes to the Prices and the Subcontract Completion Date from the quotation which he has accepted or from his own assessment in his notification implementing a compensation event.

**Payment on termination** **97**    The amount due on termination is assessed without taking grouping of activities into account.
            97.3   vities into account.

# Option B: Priced subcontract with bill of quantities

| | | |
|---|---|---|
| **Identified and defined terms** | **11** | |
| | 11.2 | (18) The Fee is the amount calculated by applying the *fee percentage* to the amount of Actual Cost. |

(21) The Prices are the lump sums and the amounts obtained by multiplying the rates by the quantities for the items in the *bill of quantities* unless later changed in accordance with this subcontract.

(25) The Price for Work Done to Date is the total of

- the quantity of the work which the *Subcontractor* has completed for each item in the *bill of quantities* multiplied by the rate and
- a proportion of each lump sum which is the proportion of the work covered by the item which the *Subcontractor* has completed.

(28) Actual Cost is the cost of the components in the Schedule of Cost Components whether work is subsubcontracted or not.

**Acceleration**     **36**

36.3    When the *Contractor* accepts a quotation for an acceleration, he changes the Subcontract Completion Date and the Prices accordingly and accepts the revised programme.

**The *bill of quantities***     **55**

55.1    Information in the *bill of quantities* is not Subcontract Works Information or Site Information.

**Compensation events**     **60**

60.4    A difference between the final total quantity of work done and the quantity stated for an item in the *bill of quantities* at the Subcontract Date is a compensation event if

- the difference causes the Actual Cost per unit of quantity to change and
- the rate in the *bill of quantities* for the item at the Subcontract Date multiplied by the quantity of work done is more than 0.1% of the total of the Prices at the Subcontract Date.

If the Actual Cost per unit of quantity is reduced, the affected rate is reduced.

60.5    A difference between the final total quantity of work done and the quantity for an item stated in the *bill of quantities* at the Subcontract Date which delays Completion is a compensation event.

60.6    The *Contractor* corrects mistakes in the *bill of quantities* which are departures from the *method of measurement* or are due to ambiguities or inconsistencies. Each such correction is a compensation event which may lead to reduced Prices.

**Assessing compensation events**     **63**

63.9    Assessments for changed Prices for compensation events are in the form of changes to the *bill of quantities*.

63.10    Assessment of a compensation event for work which includes subsubcontracted work does not include fees paid or to be paid by the *Subcontractor* to Subsubcontractors.

63.11    The *Contractor* may instruct the *Subcontractor* to assess the effect of a compensation event upon Actual Cost using the short method set out in the

Schedule of Cost Components if the *Subcontractor* agrees and may make his own assessments using the short method.

**Implementing**    **65**
**compensation events**    65.4    The *Contractor* includes the changes to the Prices and the Subcontract Completion Date from the quotation which he has accepted or from his own assessment in his notification implementing a compensation event.

# Option C: Target subcontract with activity schedule

**Identified and defined**    **11**
**terms**    11.2    (19) The Fee is the amount calculated by applying the *fee percentage* to the amount of Actual Cost after first deducting any Disallowed Cost.

(20) The Prices are the lump sum prices for each of the activities in the *activity schedule* unless later changed in accordance with this subcontract.

(23) The Price for Work Done to Date is the Actual Cost which the *Subcontractor* has paid, less any Disallowed Cost, plus the Fee.

(27) Actual Cost is the amount of payments due to Subsubcontractors for work which is subsubcontracted and the cost of the components in the Schedule of Cost Components for work which is not subsubcontracted.

(30) Disallowed Cost is cost which the *Contractor* decides

- is not justified by the *Subcontractor*'s accounts and records,
- should not have been paid to a subsubcontractor in accordance with his subsubcontract,
- was incurred only because the *Subcontractor* did not

  - follow an acceptance or procurement procedure stated in the Subcontract Works Information or
  - give an early warning which he could have given or
  - results from paying a Subsubcontractor more for a compensation event than is included in the accepted quotation or assessment for the compensation event

and the cost of

- correcting faulty work after Completion,
- correcting Defects caused by the *Subcontractor* not complying with a requirement for how he is to Provide the Subcontract Works stated in the Subcontract Works Information,
- Plant and Materials not used to Provide the Subcontract Works (after allowing for reasonable wastage) and
- Equipment and people not used to Provide the Subcontract Works (after allowing for reasonable availability and utilisation) or not taken away from the Working Areas when the *Contractor* requested.

**Providing the**    **20**
**Subcontract Works**    20.3    The *Subcontractor* advises the *Contractor* on the practical implications of the design of the *subcontract works* and on subsubcontracting arrangements.

20.4    The *Subcontractor* prepares forecasts of the total Actual Cost for the *subcontract works* in consultation with the *Contractor* and submits them to the *Contractor*. Forecasts are prepared at the intervals stated in the Subcontract Data from the *subcontract starting date* until Completion of the whole of the *subcontract works*. An explanation of the changes made since the previous forecast is submitted with each forecast.

**Subsubcontracting**    **26**
26.3    The *Subcontractor* submits the proposed conditions of contract and contract data for a subsubcontract to the *Contractor* for acceptance if the *Contractor* instructs him to. A reason for not accepting is that the proposed conditions or contract data are not compatible with the obligations of the *Subcontractor* under this subcontract.

| | | |
|---|---|---|
| **The Accepted Programme** | **31** | |
| | 31.4 | The start and finish of each activity on the *activity schedule* are shown on the Accepted Programme. |
| **Acceleration** | **36** | |
| | 36.3 | When the *Contractor* accepts a quotation for an acceleration, he changes the Subcontract Completion Date and the Prices accordingly and accepts the revised programme. |
| | 36.5 | The *Subcontractor* submits a Subsubcontractor's proposal to accelerate to the *Contractor* for acceptance. |
| **Assessing the amount due** | **50** | |
| | 50.6 | Payments of Actual Cost made by the *Subcontractor* in a currency other than the *currency of this subcontract* are included in the amount due as payments to be made to him in the same currency. Such payments are converted to the *currency of this subcontract* in order to calculate the Fee and any *Subcontractor*'s share using the *exchange rates*. |
| **Actual Cost** | **52** | |
| | 52.2 | The *Subcontractor* keeps |

- accounts of his payments of Actual Cost,
- records which show that the payments have been made,
- records of communications and calculations relating to assessment of compensation events for Subsubcontractors and
- other accounts and records as stated in the Subcontract Works Information.

| | | |
|---|---|---|
| | 52.3 | The *Subcontractor* allows the *Contractor* to inspect at any time within working hours the accounts and records which he is required to keep. |
| **The *Subcontractor*'s share** | **53** | |
| | 53.1 | The *Subcontractor*'s share of the difference between the total of the Prices and the Price for Work Done to Date is assessed using the percentages stated in the Subcontract Data. The *Contractor* assesses the *Subcontractor*'s share by applying each percentage to that proportion of the difference between the Price for Work Done to Date and the total of the Prices which is within each range stated in the Subcontract Data. |
| | 53.2 | If the Price for Work Done to Date is less than the total of the Prices, the *Subcontractor* is paid his share of the saving. If the Price for Work Done to Date is greater than the total of the Prices, the *Subcontractor* pays his share of the excess. |
| | 53.3 | The *Contractor* assesses the *Subcontractor*'s share at Completion of the whole of the *subcontract works* using his forecasts of the final Price for Work Done to Date and the final total of the Prices. This share is included in the amount due following Completion of the whole of the *subcontract works*. |
| | 53.4 | The *Contractor* again assesses the *Subcontractor*'s share using the final Price for Work Done to Date and the final total of the Prices. This share is included in the final amount due. |
| | 53.5 | If the *Contractor* accepts a proposal by the *Subcontractor* to change the Subcontract Works Information provided by the *Contractor* so that the Actual Cost is reduced the Prices are not reduced. |
| **The *activity schedule*** | **54** | |
| | 54.1 | Information in the *activity schedule* is not Subcontract Works Information or Site Information. |

54.2     If the *Subcontractor* changes a planned method of working at his discretion so that the *activity schedule* does not comply with the Accepted Programme, the *Subcontractor* submits a revision of the *activity schedule* to the *Contractor* for acceptance.

54.3     Reasons for not accepting a revision of the *activity schedule* are that

- it does not comply with the Accepted Programme,
- any changed Prices are not reasonably distributed between the activities and
- the total of the Prices is changed.

**Assessing compensation events**    **63**

63.8     Assessments for changed Prices for compensation events are in the form of changes to the *activity schedule*.

63.11     The *Contractor* may instruct the *Subcontractor* to assess the effect of a compensation event upon Actual Cost using the short method set out in the Schedule of Cost Components if the *Subcontractor* agrees and may make his own assessments using the short method.

**Implementing compensation events**    **65**

65.4     The *Contractor* includes the changes to the Prices and the Completion Date from the quotation which he has accepted or from his own assessment in his notification implementing a compensation event.

**Payment on termination**    **97**

97.4     If there is a termination, the *Contractor* assesses the *Subcontractor*'s share after he has certified termination. His assessment uses the Price for Work Done to Date at termination and the total of the Prices for the work done before termination.

# Option D: Target subcontract with bill of quantities

**Identified and defined** **11**

**terms** 11.2    (19) The Fee is the amount calculated by applying the *fee percentage* to the amount of Actual Cost after first deducting any Disallowed Cost.

(21) The Prices are the lump sums and the amounts obtained by multiplying the rates by the quantities for the items in the *bill of quantities* unless later changed in accordance with this subcontract.

(23) The Price for Work Done to Date is the Actual Cost which the *Subcontractor* has paid, less any Disallowed Cost, plus the Fee.

(27) Actual Cost is the payments due to Subsubcontractors for work which is subsubcontracted and the cost of the components in the Schedule of Cost Components for work which is not subsubcontracted.

(30) Disallowed Cost is cost which the *Contractor* decides

- is not justified by the *Subcontractor*'s accounts and records,
- should not have been paid to a Subsubcontractor in accordance with his subsubcontract,
- was incurred only because the *Subcontractor* did not

  - follow an acceptance or procurement procedure stated in the Subcontract Works Information or
  - give an early warning which he could have given or

- results from paying a Subsubcontractor more for a compensation event than is included in the accepted quotation or assessment for the compensation event

and the cost of

- correcting faulty work after Completion,
- correcting Defects caused by the *Subcontractor* not complying with a requirement for how he is to Provide the Subcontract Works stated in the Subcontract Works Information,
- Plant and Materials not used to Provide the Subcontract Works (after allowing for reasonable wastage) and
- Equipment and people not used to Provide the Subcontract Works (after allowing for reasonable availability and utilisation) or not taken away from the Working Areas when the *Contractor* requested.

**Providing the** **20**

**Subcontract Works** 20.3    The *Subcontractor* advises the *Contractor* on the practical implications of the design of the subcontract works and on subsubcontracting arrangements.

20.4    The *Subcontractor* prepares forecasts of the total Actual Cost for the *subcontract works* in consultation with the *Contractor* and submits them to the *Contractor*. Forecasts are prepared at the intervals stated in the Subcontract Data from the *subcontract starting date* until Completion of the whole of the *subcontract works*. An explanation of the changes made since the previous forecast is submitted with each forecast.

**Subsubcontracting** **26**

26.3    The *Subcontractor* submits the proposed conditions of subcontract and subcontract data for a subsubcontract to the *Contractor* for acceptance if the *Contractor* instructs him to. A reason for not accepting is that the proposed conditions or subcontract data are not compatible with the obligations of the *Subcontractor* under this subcontract.

| | | |
|---|---|---|
| **Acceleration** | **36** | |
| | 36.3 | When the *Contractor* accepts a quotation for an acceleration, he changes the Subcontract Completion Date and the Prices accordingly and accepts the revised programme. |
| | 36.5 | The *Subcontractor* submits a Subsubcontractor's proposal to accelerate to the *Contractor* for acceptance. |

| | | |
|---|---|---|
| **Assessing the amount due** | **50** | |
| | 50.6 | Payments of Actual Cost made by the *Subcontractor* in a currency other than the *currency of this subcontract* are included in the amount due as payments to be made to him in the same currency. Such payments are converted to the *currency of this subcontract* in order to calculate the Fee and any *Subcontractor*'s share using the *exchange rates*. |

| | | |
|---|---|---|
| **Actual Cost** | **52** | |
| | 52.2 | The *Subcontractor* keeps |

- accounts of his payments of Actual Cost,
- records which show that the payments have been made,
- records of communications and calculations relating to assessment of compensation events for Subsubcontractors and
- other accounts and records as stated in the Subcontract Works Information.

| | | |
|---|---|---|
| | 52.3 | The *Subcontractor* allows the *Contractor* to inspect at any time within working hours the accounts and records which he is required to keep. |

| | | |
|---|---|---|
| **The *Subcontractor*'s share** | **53** | |
| | 53.1 | The *Subcontractor*'s share of the difference between the total of the Prices and the Price for Work Done to Date is assessed using the percentages stated in the Subcontract Data. The *Contractor* assesses the *Subcontractor*'s share by applying each percentage to that proportion of the difference between the Price for Work Done to Date and the total of the Prices which is within each range stated in the Subcontract Data. |
| | 53.2 | If the Price for Work Done to Date is less than the total of the Prices, the *Subcontractor* is paid his share of the saving. If the Price for Work Done to Date is greater than the total of the Prices, the *Subcontractor* pays his share of the excess. |
| | 53.3 | The *Contractor* assesses the *Subcontractor*'s share at Completion of the whole of the *subcontract works* using his forecasts of the final Price for Work Done to Date and the final total of the Prices. This share is included in the amount due following Completion of the whole of the *subcontract works*. |
| | 53.4 | The *Contractor* again assesses the *Subcontractor*'s share using the final Price for Work Done to Date and the final total of the Prices. This share is included in the final amount due. |
| | 53.5 | If the *Contractor* accepts a proposal by the *Subcontractor* to change the Subcontract Works Information provided by the *Contractor* which will reduce Actual Cost, the Prices are not reduced. |

| | | |
|---|---|---|
| **The *bill of quantities*** | **55** | |
| | 55.1 | Information in the *bill of quantities* is not Subcontract Works Information or Site Information. |

| | | |
|---|---|---|
| **Compensation events** | **60** | |
| | 60.4 | A difference between the final total quantity of work done and a quantity for an item stated in the *bill of quantities* at the Subcontract Date is a compensation event if |

- the difference causes the Actual Cost per unit of quantity to change and
- the rate in the *bill of quantities* for the item at the Subcontract Date multiplied by the quantity of work done is more than 0.1% of the total of the Prices at the Subcontract Date.

If the Actual Cost per unit of quantity is reduced, the affected rate is reduced.

60.5 A difference between the final total quantity of work done and the quantity for an item stated in the *bill of quantities* at the Subcontract Date which delays Completion is a compensation event.

60. 6 The *Contractor* corrects mistakes in the *bill of quantities* which are departures from the *method of measurement* or are due to ambiguities or inconsistencies. Each such correction is a compensation event which may lead to reduced Prices.

**Assessing compensation events** **63**

63.9 Assessments for changed Prices for compensation events are in the form of changes to the *bill of quantities*.

63.11 The *Contractor* may instruct the *Subcontractor* to assess the effect of a compensation event upon Actual Cost using the short method set out in the Schedule of Cost Components if the *Subcontractor* agrees and may make his own assessments using the short method.

**Implementing compensation events** **65**

65.4 The *Contractor* includes the changes to the Prices and the Subcontract Completion Date from the quotation which he has accepted or from his own assessment in his notification implementing a compensation event.

**Payment on termination** **97**

97.4 If there is a termination, the *Contractor* assesses the *Subcontractor*'s share after he has certified termination. His assessment uses the Price for Work Done to Date at termination and the total of the Prices for the work done before termination.

# Option E: Cost reimbursable subcontract

| | |
|---|---|
| **Identified and defined** **11** | |
| **terms** 11.2 | (19) The Fee is the amount calculated by applying the *fee percentage* to the amount of Actual Cost after first deducting any Disallowed Cost. |

(22) The Prices are the Actual Cost less Disallowed Cost plus the Fee.

(23) the Price for Work Done to Date is the Actual Cost which the *Subcontractor* has paid, less any Disallowed Cost, plus the Fee.

(27) Actual Cost is the payments due to Subsubcontractors for work which is subsubcontracted and the cost of the components in the Schedule of Cost Components for work which is not subsubcontracted.

(30) Disallowed Cost is cost which the *Contractor* decides

- is not justified by the *Subcontractor*'s accounts and records,
- should not have been paid to a Subsubcontractor in accordance with his subsubcontract,
- was incurred only because the *Subcontractor* did not

    - follow an acceptance or procurement procedure stated in the Subcontract Works Information or
    - give an early warning which he could have given or

- results from paying a Subsubcontractor more for a compensation event than is included in the accepted quotation or assessment for the compensation event

and the cost of

- correcting faulty work after Completion,
- correcting Defects caused by the *Subcontractor* not complying with a requirement for how he is to Provide the Subcontract Works stated in the Subcontract Works Information,
- Plant and Materials not used to Provide the Subcontract Works (after allowing for reasonable wastage) and
- Equipment and people not used to Provide the Subcontract Works (after allowing for reasonable availability and utilisation) or not taken away from the Working Areas when the *Contractor* requested.

| | |
|---|---|
| **Providing the** **20** | |
| **Subcontract Works** 20.3 | The *Subcontractor* advises the *Contractor* on the practical implications of the design of the *subcontract works* and on subsubcontracting arrangements. |
| 20.4 | The *Subcontractor* prepares forecasts of the total Actual Cost for the *subcontract works* in consultation with the *Contractor* and submits them to the *Contractor*. Forecasts are prepared at the intervals stated in the Subcontract Data from the *subcontract starting date* until Completion of the whole of the *subcontract works*. An explanation of the changes made since the previous forecast is submitted with each forecast. |
| **Subsubcontracting** **26** | |
| 26.3 | The *Subcontractor* submits the proposed conditions of subcontract and subcontract data for a subsubcontract to the *Contractor* for acceptance if the *Contractor* instructs him to. A reason for not accepting is that the proposed conditions or subcontract data are not compatible with the obligations of the *Subcontractor* under this subcontract. |

**Acceleration** **36**

36.4 When the *Contractor* accepts a quotation for an acceleration, he changes the Subcontract Completion Date accordingly and accepts the revised programme.

36.5 The *Subcontractor* submits a Subsubcontractor's proposal to accelerate to the *Contractor* for acceptance.

**Assessing the amount due** **50**

50.7 Payments of Actual Cost made by the *Subcontractor* in a currency other than the *currency of this subcontract* are included in the amount due as payments to be made to him in the same currency. Such payments are converted to the *currency of this subcontract* in order to calculate the Fee at the *exchange rates*.

**Actual Cost** **52**

52.2 The *Subcontractor* keeps

- accounts of his payments of Actual Cost,
- records which show that the payments have been made,
- records of communications and calculations relating to assessment of compensation events for Subsubcontractors and
- other accounts and records as stated in the Subcontract Works Information.

52.3 The *Subcontractor* allows the *Contractor* to inspect at any time within working hours the accounts and records which he is required to keep.

**Assessing compensation events** **63**

63.11 The *Contractor* may instruct the *Subcontractor* to assess the effect of a compensation event upon Actual Cost using the short method set out in the Schedule of Cost Components if the *Subcontractor* agrees and may make his own assessments using the short method.

**Implementing compensation events** **65**

65.3 The *Contractor* includes the changes to the forecast amount of the Prices and the Subcontract Completion Date in his notification to the *Subcontractor* implementing a compensation event.

65.5 The *Subcontractor* does not implement a subsubcontract compensation event until it has been agreed by the *Contractor*.

# SECONDARY OPTION CLAUSES

## Option G: Performance bond

**Performance bond**   **G1**

G1.1    A bank or insurer which the *Contractor* has accepted gives the *Contractor* a performance bond in the form set out in the Subcontract Works Information and for the amount stated in the Subcontract Data. If the bond was not given by the Subcontract Date, it is given to the *Contractor* within four weeks of the Subcontract Date. A reason for not accepting the bank or insurer is that its commercial position is not strong enough to carry the bond.

## Option H: Parent company guarantee

**Parent company**   **H1**
**guarantee**   H1.1    If a parent company owns the *Subcontractor*, the parent company gives a guarantee of the *Subcontractor*'s performance to the *Contractor* in the form set out in the Subcontract Works Information. If it was not given by the Subcontract Date, it is given within four weeks of the Subcontract Date.

## Option J : Advanced payment to the *Subcontractor*

**Advanced payment**   **J1**
J1.1    The *Contractor* makes an advanced payment to the *Subcontractor* of the amount stated in the Subcontract Data.

J1.2    The advanced payment is made either within five weeks of the Subcontract Date or, if an advanced payment bond is required but has not been given before the Subcontract Date, within five weeks of a bank or insurer which the *Contractor* has accepted giving an advanced payment bond to the *Contractor*. A reason for not accepting the proposed bank or insurer is that its commercial position is not strong enough to carry the bond. The bond is for the amount of the advanced payment and in the form set out in the Subcontract Works Information. Delayed payment of the advanced payment is a compensation event.

J1.3    The advanced payment is repaid to the *Contractor* by the *Subcontractor* in instalments of the amount stated in the Subcontract Data. An additional instalment is included in each amount due assessed after the period stated in the Subcontract Data has passed until the advanced payment has been repaid.

# Option K: Multiple currencies (used only with Options A and B)

**Multiple currencies** **K1**

K1.1    The *Subcontractor* is paid in currencies other than the *currency of this subcontract* for the work listed in the Subcontract Data. The *exchange rates* are used to convert from the *currency of this subcontract* to other currencies.

K1.2    Payments to the *Subcontractor* in currencies other than *the currency of this subcontract* do not exceed the maximum amounts stated in the Subcontract Data. Any excess is paid in the *currency of this subcontract.*

# Option L: Sectional Completion

**Sectional Completion** **L1**

L1.1    Each reference in the *conditions of subcontract* to the *subcontract works*, to Completion and to the Subcontract Completion Date applies to the *subcontract works* and any *section* of the *subcontract works* unless it is stated to apply to the whole of the *subcontract works*.

# Option M: Limitation of the *Subcontractor*'s liability for his design to reasonable skill and care

**The *Subcontractor*'s** **M1**
**design**   M1.1    The *Subcontractor* is not liable for Defects in the *subcontract works* due to his design so far as he proves that he used reasonable skill and care to ensure that it complied with the Subcontract Works Information.

# Option N: Price adjustment for inflation (used only with Options A, B, C and D)

**Defined terms** **N1**

N1.1    (a) The Base Date Index (B) is the latest available index before the *base date*

(b) The Latest Index (L) is the latest available index before the date of assessment of an amount due.

(c) The Price Adjustment Factor is the total of the products of each of the proportions stated in the Subcontract Data multiplied by $(L - B)/B$ for the index linked to it.

**Price Adjustment Factors** **N2**

N2.1    If an index is changed after it has been used in calculating a Price Adjustment

Factor, the calculation is repeated and a correction included in the next assessment of the amount due.

N2.2    The Price Adjustment Factor calculated at the Subcontract Completion Date for the whole of the *subcontract works* is used for calculating the price adjustment after this date.

**Compensation events**   **N3**

N3.1    The Actual Cost for compensation events is assessed using the

- Actual Costs current at the time of assessing the compensation event adjusted to *base date* by dividing by one plus the Price Adjustment Factor for the last assessment of the amount due and
- Actual Costs at *base date* levels for amounts calculated from rates stated in the Subcontract Data for employees and Equipment.

**Price adjustment**   **N4**

Options A and B   N4.1    Each amount due includes an amount for price adjustment which is the sum of

- the change in the Price for Work Done to Date since the last assessment of the amount due multiplied by the Price Adjustment Factor for the date of the current assessment,
- the amount for price adjustment included in the previous amount due and
- correcting amounts, not included elsewhere, which arise from changes to indices used for assessing previous amounts for price adjustment.

Options C and D   N4.2    Each time the amount due is assessed, an amount for price adjustment is added to the total of the Prices which is the sum of

- the change in the Price of Work Done to Date since the last assessment of the amount due multiplied by $(1 - 1/(1 + PAF))$ where PAF is the Price Adjustment Factor for the date of the current assessment and
- correcting amounts, not included elsewhere, which arise from changes to indices used for assessing previous amounts for price adjustment.

# Option P: Retention

**Retention**   **P1**

P1.1    Nothing is retained until the Price for Work Done to Date has reached the *retention free amount*. An amount is then retained from the *Subcontractor* in each amount due until the earlier of Completion of the whole of the *subcontract works* and the *Contractor* taking over the whole of the *subcontract works*. This amount is the *retention percentage* applied to the excess of the Price for Work Done to Date above the *retention free amount*.

P1.2    The amount retained is halved in the amount due assessed at Completion or in the first assessment after the *Contractor* has taken over the whole of the *subcontract works* if this is before Completion of the whole of the *subcontract works*. The amount retained remains at this amount until the Defects Certificate is issued. Nothing is retained in the assessment made when the Defects Certificate is issued or in later assessments of the amount due.

## Option Q: Bonus for early Completion

**Bonus for early** **Q1**
**Completion** Q1.1 The *Subcontractor* is paid a bonus calculated at the rate stated in the Subcontract Data for each day from the earlier of Completion and the date on which the *Employer* begins to use the *subcontract works* until the Subcontract Completion Date.

## Option R: Delay damages

**Delay damages** **R1**
R1.1 The *Subcontractor* pays delay damages at the rate stated in the Subcontract Data for each day from the Subcontract Completion Date until the earlier of Completion and the date on which the *Employer* begins to use the *subcontract works*.

R1.2 If the Subcontract Completion Date is delayed after delay damages have been paid, the *Contractor* repays the overpayment of damages with interest. Interest is assessed from the date of payment to the date of repayment and the date of repayment is an assessment date.

## Option S: Low performance damages

**Low performance** **S1**
**damages** S1.1 If a Defect included in the Defects Certificate shows low performance with respect to a performance level stated in the Subcontract Data, the *Subcontractor* pays the amount of low performance damages stated in the Subcontract Data.

## Option T: Changes in the law

**Changes in the law** **T1**
T1.1 A change in the law of the country in which the Site is located is a compensation event if it occurs after the Subcontract Date. The *Contractor* may notify the *Subcontractor* of a compensation event for a change in the law and instruct him to submit quotations. If the effect of a compensation event which is a change in the law is to reduce the total Actual Cost, the Prices are reduced.

# Option U:  Special conditions of subcontract

**Special conditions of** **U1**
**subcontract**   U1.1   The special conditions of subcontract stated in the Subcontract Data are part of this subcontract.

# SCHEDULE OF COST COMPONENTS

When Option C, D or E is used, in this schedule the *Subcontractor* means the *Subcontractor* and not his Subsubcontractors. Amounts are included only in one cost component.

**People 1** The following components of the cost of

- people who are directly employed by the *Subcontractor* and either working within the Working Areas or whose normal place of working is within the Working Areas and
- people who are not directly employed by the *Subcontractor* but are paid by the *Subcontractor* according to the time worked while they are within the Working Areas.

11 Wages and salaries.

12 Payments to people for

(a) bonuses and incentives
(b) overtime
(c) working in special circumstances
(d) special allowances
(e) absence due to sickness and holidays
(f) severance related to work on this subcontract.

13 Payments made in relation to people for

(a) travelling to and from the Working Areas
(b) subsistence and lodging
(c) relocation
(d) medical examinations
(e) passports and visas
(f) travel insurance
(g) items (a) to (f) for a spouse or dependents
(h) protective clothing
(j) meeting the requirements of the law
(k) superannuation and life assurance
(l) death benefit
(m) occupational accident benefits
(n) medical aid.

**Equipment 2** The following components of the cost of Equipment which is used within the Working Areas (excluding Equipment cost covered by the percentage for Working Areas overheads)

21 Payments for hire of Equipment not owned by the *Subcontractor*, by the *Subcontractor*'s parent company or by another part of a group with the same parent company.

22 An amount for depreciation and maintenance of Equipment which is

(a) owned by the *Subcontractor*,
(b) purchased by the *Subcontractor* under a hire purchase or lease agreement or
(c) hired by the *Subcontractor* from the *Subcontractor*'s parent company or another part of a group with the same parent company.

The depreciation and maintenance charge is the actual purchase price of the item of Equipment (or first cost if the *Subcontractor* assembled the item) divided by its average working life remaining at the time of purchase (expressed in weeks).

The amount for depreciation and maintenance is calculated by multiplying the depreciation and maintenance charge by the time required expressed as a number of weeks) and then adding the percentage for Equipment depreciation and maintenance stated in the Subcontract Data.

23    The purchase price of Equipment which is consumed.

24    Except when covered by hire rates, payments for

(a) transporting Equipment to and from the Working Areas and
(b) erecting and dismantling Equipment.

**Plant and Materials**  **3**    The following components of the cost of Plant and Materials

31    Payments for

(a) purchasing Plant and Materials
(b) delivery to and removal from the Working Areas
(c) providing and removing packaging
(d) samples and tests.

32    Cost is credited with payments received for disposal of Plant and Materials.

**Charges**  **4**    The following components of the cost of charges paid by the *Subcontractor*

41    Payments to utilities for provision and use in the Working Areas of

(a) water
(b) gas
(c) electricity
(d) other services.

42    Payments to public authorities, utilities and other properly constituted authorities of charges which they are authorised to make in respect of the *subcontract works*.

43    Payments for

(a) financing charges (excluding charges compensated for by interest paid in accordance with this subcontract)
(b) buying or leasing land
(c) compensation for loss of crops or buildings
(d) royalties
(e) inspection certificates
(f) rent of premises in the Working Areas
(g) charges for access to the Working Areas
(h) facilities for visits to the Working Areas by Others
(i) specialist services.

44    A charge for overhead costs incurred within the Working Areas calculated by applying the percentage for Working Areas overheads stated in the Subcontract Data to the Actual Cost of people. The charge is deemed to include provision and use of accommodation, equipment, supplies and services for

(a) offices and drawing offices
(b) laboratories
(c) workshops
(d) stores and compounds
(e) labour camps
(f) cabins
(g) catering

(h) medical facilities and first aid

(j) recreation

(k) sanitation

(l) security

(m) copying

(n) telephone, telex, fax, radio and CCTV

(o) surveying and setting out

(p) computing

(q) tools.

**Manufacture and fabrication** 5 The following components of the cost of manufacture and fabrication done by the *Subcontractor* or a Subsubcontractor outside the Working Areas

51 The total of the hours worked by employees multiplied by the hourly rates stated in the Subcontract Data for the categories of employees listed and

52 An amount for overheads calculated by multiplying this total by the percentage for manufacturing and fabrication overheads stated in the Subcontract Data.

**Design** 6 The following components of the cost of design of the *subcontract works* and Equipment done outside the Working Areas

61 The total of the hours worked by employees multiplied by the hourly rates stated in the Subcontract Data for the categories of employees listed and

62 An amount for overheads calculated by multiplying this total by the percentage for design overheads stated in the Subcontract Data and

63 The cost of travel to and from the Working Areas for the categories of people listed in the Subcontract Data.

**Insurance** 7 Insurer's payments of claims are deducted from cost.

# SCHEDULE OF COST COMPONENTS USING THE SHORT METHOD

When Option C, D or E is used, in this schedule the *Subcontractor* means the *Subcontractor* and not his Subsubcontractors. Amounts are included only in one cost component.

**People 1** The following components of the cost of

- people who are directly employed by the *Subcontractor* and either working within the Working Areas or whose normal place of work is within the Working Areas and
- people who are not directly employed by the *Subcontractor* but paid by the *Subcontractor* according to the time worked whilst they are within the Working Areas.

11 Wages and salaries

12 Payments to people for

(a) bonuses and incentives
(b) overtime
(c) working in special circumstances
(d) special allowances

13 Payments made in relation to people for

(a) travelling to and from the Working Areas
(b) subsistence and lodging

14 A charge for overhead costs for people, payments to Others and Working Areas overheads calculated by applying the percentage for people overheads stated in the Subcontract Data to the total of items 11, 12 and 13.

**Equipment 2** The following components of the cost of Equipment used within the Working Areas (excluding Equipment cost covered by the percentage for people overheads).

21 Amounts for Equipment which is in the published list stated in the Subcontract Data. These amounts are calculated by applying the percentage adjustment for listed Equipment stated in the Subcontract Data to the rates in the published list and by multiplying the resulting rate by the time for which the Equipment is working.

22 Amounts for Equipment which is not in the published list stated in the Subcontract Data. These amounts are calculated by multiplying the rates stated for each item listed in the Subcontract Data multiplied by the time for which the Equipment is working.

**Plant and Materials 3** The following components of the cost of Plant and Materials

31 Payments for

(a) purchasing Plant and Materials
(b) delivery to and removal from the Working Areas
(c) providing and removing packaging
(d) samples and tests.

32 Cost is credited with payments received for disposal of Plant and Materials.

| | | |
|---|---|---|
| **Manufacture and fabrication** | 5 | The following components of the cost of manufacture and fabrication done by the *Subcontractor* or a Subsubcontractor outside the Working Areas |
| | 51 | The total of the hours worked by employees multiplied by the hourly rates stated in the Subcontract Data for the categories of employees listed and |
| | 52 | An amount for overheads calculated by multiplying this total by the percentage for manufacturing and fabrication overheads stated in the Subcontract Data. |
| **Design** | 6 | The following components of the cost of design of the *subcontract works* and Equipment done outside the Working Areas |
| | 61 | The total of the hours worked by employees multiplied by the hourly rates stated in the Subcontract Data for the categories of employees listed and |
| | 62 | An amount for overheads calculated by multiplying this total by the percentage for design overheads stated in the Subcontract Data and |
| | 63 | The cost of travel to and from the Working Areas for the categories of people listed in the Subcontract Data. |
| **Insurance** | 7 | Insurer's payments of claims are deducted from cost. |

# SUBCONTRACT DATA

## Part one – Data provided by the Contractor

**Statements given in all subcontracts**

1. General The *conditions of subcontract* are the core clauses and the clauses for Options . . . . . . of the New Engineering Subcontract.

- The *works* in the main contract are

  . . . . . . . . . . . . . . . . . . . . . . . . . . . . . . . . . . . . . . . . . . . . . . . . . . . . . . . . . . . . . . . . . . . . . . . . .

- The *subcontract works* are

  . . . . . . . . . . . . . . . . . . . . . . . . . . . . . . . . . . . . . . . . . . . . . . . . . . . . . . . . . . . . . . . . . . . . . . . . .

- The *Employer* in the main contract is

  Name . . . . . . . . . . . . . . . . . . . . . . . . . . . . . . . . . . . . . . . . . . . . . . . . . . . . . . . . . . . . . . . . . . . . . .

  Address . . . . . . . . . . . . . . . . . . . . . . . . . . . . . . . . . . . . . . . . . . . . . . . . . . . . . . . . . . . . . . . . . . . .

  . . . . . . . . . . . . . . . . . . . . . . . . . . . . . . . . . . . . . . . . . . . . . . . . . . . . . . . . . . . . . . . . . . . . . . . . .

- The *Adjudicator* in this subcontract is

  Name . . . . . . . . . . . . . . . . . . . . . . . . . . . . . . . . . . . . . . . . . . . . . . . . . . . . . . . . . . . . . . . . . . . . . .

  Address . . . . . . . . . . . . . . . . . . . . . . . . . . . . . . . . . . . . . . . . . . . . . . . . . . . . . . . . . . . . . . . . . . . .

  . . . . . . . . . . . . . . . . . . . . . . . . . . . . . . . . . . . . . . . . . . . . . . . . . . . . . . . . . . . . . . . . . . . . . . . . .

- The Subcontract Works Information is in

  . . . . . . . . . . . . . . . . . . . . . . . . . . . . . . . . . . . . . . . . . . . . . . . . . . . . . . . . . . . . . . . . . . . . . . . . .

  . . . . . . . . . . . . . . . . . . . . . . . . . . . . . . . . . . . . . . . . . . . . . . . . . . . . . . . . . . . . . . . . . . . . . . . . .

  . . . . . . . . . . . . . . . . . . . . . . . . . . . . . . . . . . . . . . . . . . . . . . . . . . . . . . . . . . . . . . . . . . . . . . . . .

  . . . . . . . . . . . . . . . . . . . . . . . . . . . . . . . . . . . . . . . . . . . . . . . . . . . . . . . . . . . . . . . . . . . . . . . . .

  . . . . . . . . . . . . . . . . . . . . . . . . . . . . . . . . . . . . . . . . . . . . . . . . . . . . . . . . . . . . . . . . . . . . . . . . .

  . . . . . . . . . . . . . . . . . . . . . . . . . . . . . . . . . . . . . . . . . . . . . . . . . . . . . . . . . . . . . . . . . . . . . . . . .

  . . . . . . . . . . . . . . . . . . . . . . . . . . . . . . . . . . . . . . . . . . . . . . . . . . . . . . . . . . . . . . . . . . . . . . . . .

- The Site Information is in

  . . . . . . . . . . . . . . . . . . . . . . . . . . . . . . . . . . . . . . . . . . . . . . . . . . . . . . . . . . . . . . . . . . . . . . . . .

  . . . . . . . . . . . . . . . . . . . . . . . . . . . . . . . . . . . . . . . . . . . . . . . . . . . . . . . . . . . . . . . . . . . . . . . . .

  . . . . . . . . . . . . . . . . . . . . . . . . . . . . . . . . . . . . . . . . . . . . . . . . . . . . . . . . . . . . . . . . . . . . . . . . .

  . . . . . . . . . . . . . . . . . . . . . . . . . . . . . . . . . . . . . . . . . . . . . . . . . . . . . . . . . . . . . . . . . . . . . . . . .

  . . . . . . . . . . . . . . . . . . . . . . . . . . . . . . . . . . . . . . . . . . . . . . . . . . . . . . . . . . . . . . . . . . . . . . . . .

  . . . . . . . . . . . . . . . . . . . . . . . . . . . . . . . . . . . . . . . . . . . . . . . . . . . . . . . . . . . . . . . . . . . . . . . . .

  . . . . . . . . . . . . . . . . . . . . . . . . . . . . . . . . . . . . . . . . . . . . . . . . . . . . . . . . . . . . . . . . . . . . . . . . .

- The *boundaries of the site* are . . . . . . . . . . . . . . . . . . . . . . . . . . . . . . . . . . . . . . . . .

- The *language of this subcontract* is . . . . . . . . . . . . . . . . . . . . . . . . . . . . . . . . . . . . .

- This subcontract is governed by the law of . . . . . . . . . . . . . . . . . . . . . . . . . . . . . . .

- The *period for reply* to a communication is

  for a reply by the *Contractor*. . . . . . . . . . . . . . . . . . . . . . . . .weeks

  for a reply by the *Subcontractor* . . . . . . . . . . . . . . . . . . . . . .weeks

**2. The *Subcontractor*'s main responsibilities**

- After Completion of the whole of the *subcontract works*, the limit of the *Subcontractor*'s liability for his design is

  . . . . . . . . . . . . . . . . . . . . . . . . . . . . . . . . . . . . . . . . . . . . . . . . . . . . . . . . . . . . . . . . . . . . . . .

**3. Time**

- The *subcontract starting date* is . . . . . . . . . . . . . . . . . . . . . . . . . . . . . . . . . . . . . . . . .

- The *subcontract possession dates* are

| Part of the Site | Date |
|---|---|
| 1 . . . . . . . . . . . . . . . | . . . . . . . . . . . . . . . . |
| 2 . . . . . . . . . . . . . . . | . . . . . . . . . . . . . . . . |
| 3 . . . . . . . . . . . . . . . | . . . . . . . . . . . . . . . . |

- The *Subcontractor* submits revised programmes at intervals no longer than

  . . . . . . . . . . . . . . . . . . . . . . . weeks

**4. Testing and Defects**

- The *defects date* is . . . . . . . . . . . . . . . . . . . . weeks after Completion of the whole of the *subcontract works*

- The *defect correction period* is . . . . . . . . . . . . . . . . . . . . . weeks

**5. Payment**

- The *currency of this subcontract* is the . . . . . . . . . . . . . . . . . . . . . . . . . . . . . . . . . .

- The *assessment interval* is . . . . . . . . . . . . . . . weeks [not more than five]

- The *interest rate* is . . . . . . . . . % per . . . . . . . . . . .(time period) above/below the

  . . . . . . . . . . . . . . . . . . . . . rate of the . . . . . . . . . . . . . . . . . . . . . . . . . . . . . . . . . .

**6. Compensation events**

- The *weather measurements* are

  - rainfall (mm). . . . . . . . . . . . . . . . . .

  - days with rainfall greater than 5 mm (nr) . . . . . . . . . . . .

  - days with minimum air temperature less than 0 degrees Celsius (nr) . . . . .

  - days with snow lying at . . . . . . . . . . . . hours GMT (nr) . . . . . . . . . . . . . .

  and these measurements

  . . . . . . . . . . . . . . . . . . . . . . . . . . . . . . . . . . . . . . . . . . . . . . . . . . . . . . . . . . . . . . . . . . . . .

  . . . . . . . . . . . . . . . . . . . . . . . . . . . . . . . . . . . . . . . . . . . . . . . . . . . . . . . . . . . . . . . . . . . . .

  . . . . . . . . . . . . . . . . . . . . . . . . . . . . . . . . . . . . . . . . . . . . . . . . . . . . . . . . . . . . . . . . . . . . .

- The *weather data* are the records of past *weather measurements* which were recorded at. . . . . . . . . . . . . . . . . . . . . . . . . . . . . . . . . . . . . . . . . . . . . . . . . . . . . . . and which are available from . . . . . . . . . . . . . . . . . . . . . . . . . . . . . . . . . . . . . . . . . .

  . . . . . . . . . . . . . . . . . . . . . . . . . . . . . . . . . . . . . . . . . . . . . . . . . . . . . . . . . . . . . . .

  Where no recorded data are available, assumed values for the ten year return monthly *weather data* are stated here.

- The place where weather is to be recorded is . . . . . . . . . . . . . . . . . . . . . . . . . . . . . .

**8. Risks and insurance**
- The maximum deductible for insurance of the *subcontract works* and of Plant and Materials is. . . . . . . . . . . . . . . . . . . . . . . . . . . . . . . . . . . . . . . . . . . . . . . . . . . . . . . . . . .

- The minimum cover for insurance of the *subcontract works* and of Plant and Materials in respect of the *Subcontractor*'s faulty design is . . . . . . . . . . . . . . . . . . . .

- The maximum deductible for insurance of Equipment is. . . . . . . . . . . . . . . . . . . . . .

- The cover for insurance of other property is . . . . . . . . . . . . . . . . . . . . . . . . . . . . . . .

- The minimum cover for insurance of other property in respect of the *Subcontractor*'s faulty design is . . . . . . . . . . . . . . . . . . . . . . . . . . . . . . . . . . . . . . . . . . . . . . . . . . . .

- The minimum cover for personal injury or death insurance
  - for the *Subcontractor*'s employees is . . . . . . . . . . . . . . . . . . . . . . . . . . . . . . .
  - and for other people is . . . . . . . . . . . . . . . . . . . . . . . . . . . . . . . . . . . . . . . . . . .

**9. Disputes and termination**
- The person who will choose an adjudicator if the Parties cannot agree a choice is

  . . . . . . . . . . . . . . . . . . . . . . . . . . . . . . . . . . . . . . . . . . . . . . . . . . . . . . . . . . . . . . .

- The *arbitration procedure* is . . . . . . . . . . . . . . . . . . . . . . . . . . . . . . . . . . . . . . . . . . .

  . . . . . . . . . . . . . . . . . . . . . . . . . . . . . . . . . . . . . . . . . . . . . . . . . . . . . . . . . . . . . . .

**Optional statements**   If the Subcontract Completion Date has been decided by the *Contractor*

- The *subcontract completion date* for the whole of the *subcontract works* is . . . . . . .

  . . . . . . . . . . . . . . . . . . . . . . . . . . . . . . . . . . . . . . . . . . . . . . . . . . . . . . . . . . . . . . .

If the *Contractor* is not willing to take over the *subcontract works* before the Subcontract Completion Date

- The *Contractor* is not willing to take over the *subcontract works* before the Sub-contract Completion Date

If the *Subcontractor* is not to submit the Accepted Programme with his tender

- The *Subcontractor* is to submit the Accepted Programme within . . . . . . . . . . . . . . weeks of the Subcontract Date

If the period for payment is not six weeks

- The period within which payments are made is . . . . . . . . . . . . . . . . . . . . . . .weeks

If there are additional compensation events

- These are compensation events

  1 . . . . . . . . . . . . . . . . . . . . . . . . . . . . . . . . . . . . . . . . . . . . . . . . . . . . . . . . . . . . . . . .

  2 . . . . . . . . . . . . . . . . . . . . . . . . . . . . . . . . . . . . . . . . . . . . . . . . . . . . . . . . . . . . . . . .

  3 . . . . . . . . . . . . . . . . . . . . . . . . . . . . . . . . . . . . . . . . . . . . . . . . . . . . . . . . . . . . . . . .

If there are additional *Employer*'s risks in the main contract

- These are *Employer*'s risks

  1 . . . . . . . . . . . . . . . . . . . . . . . . . . . . . . . . . . . . . . . . . . . . . . . . . . . . . . . . . . . . . . . .

  2 . . . . . . . . . . . . . . . . . . . . . . . . . . . . . . . . . . . . . . . . . . . . . . . . . . . . . . . . . . . . . . . .

  3 . . . . . . . . . . . . . . . . . . . . . . . . . . . . . . . . . . . . . . . . . . . . . . . . . . . . . . . . . . . . . . . .

If there are *Contractor*'s risks in the main contract which are not transferred to the *Subcontractor*

- These are *Contractor*'s risks

  1 . . . . . . . . . . . . . . . . . . . . . . . . . . . . . . . . . . . . . . . . . . . . . . . . . . . . . . . . . . . . . . . .

  2 . . . . . . . . . . . . . . . . . . . . . . . . . . . . . . . . . . . . . . . . . . . . . . . . . . . . . . . . . . . . . . . .

  3 . . . . . . . . . . . . . . . . . . . . . . . . . . . . . . . . . . . . . . . . . . . . . . . . . . . . . . . . . . . . . . . .

If the *Subcontractor* is to provide additional insurances

- The *Subcontractor* provides these insurances

  | Event | Cover |
  | --- | --- |
  | 1 . . . . . . . . . . . . . . . . | . . . . . . . . . . . . . . . . . . . . . . . . |
  | 2 . . . . . . . . . . . . . . . . | . . . . . . . . . . . . . . . . . . . . . . . . |
  | 3 . . . . . . . . . . . . . . . . | . . . . . . . . . . . . . . . . . . . . . . . . |

If the *Employer* is to provide insurances under the main contract

- The *Employer* provides these insurances

  | Event | Cover |
  | --- | --- |
  | 1 . . . . . . . . . . . . . . . . | . . . . . . . . . . . . . . . . . . . . . . . . |
  | 2 . . . . . . . . . . . . . . . . | . . . . . . . . . . . . . . . . . . . . . . . . |
  | 3 . . . . . . . . . . . . . . . . | . . . . . . . . . . . . . . . . . . . . . . . . |

If the *Contractor* is to provide insurances under this subcontract

- The *Contractor* provides these insurances

| Event | Cover |
|-------|-------|
| 1 . . . . . . . . . . . . . . . . . | . . . . . . . . . . . . . . . . . . . . . . . . |
| 2 . . . . . . . . . . . . . . . . . | . . . . . . . . . . . . . . . . . . . . . . . . |
| 3 . . . . . . . . . . . . . . . . . | . . . . . . . . . . . . . . . . . . . . . . . . |

If Option B or D is used

- The *method of measurement* is . . . . . . . . . . . . . . . . . . . . . . . . . . . . . . . . . . . . . amended as follows . . . . . . . . . . . . . . . . . . . . . . . . . . . . . . . . . . . . . . . . . . . . .
. . . . . . . . . . . . . . . . . . . . . . . . . . . . . . . . . . . . . . . . . . . . . . . . . . . . . . . . . . .
. . . . . . . . . . . . . . . . . . . . . . . . . . . . . . . . . . . . . . . . . . . . . . . . . . . . . . . . . . .

If Option C or D is used

- The *Subcontractor*'s share percentages are

| Price for Work Done to Date as a percentage of the total of the Prices | Share percentage |
|-------|-------|
| less than . . . . . . . . . . . . % | . . . . . . . . % |
| from . . . . . . % to . . . . . . % | . . . . . . . . % |
| from . . . . . . % to . . . . . . % | . . . . . . . . % |
| greater than . . . . . . . . . . . % | . . . . . . . . % |

If Option C, D or E is used

- Forecasts of Actual Cost for the *subcontract works* are prepared at intervals no longer than . . . . . . . . . . . . . . . . . . . . . . weeks

- The *exchange rates* are those published in . . . . . . . . . . . . . . . . . . . . . . . . . . . . . . . on . . . . . . . . . . . . . . . . . . . . . (date)

If Option G is used

- The amount of the performance bond is . . . . . . . . . . . . . . . . . . . . . . . . . . . . . . . . .

If Option J is used

- The amount of the advanced payment is . . . . . . . . . . . . . . . . . . . . . . . . . . . . . . . . .

- The instalments to be repaid by the *Subcontractor* in assessments starting not less than . . . . . . . . . . . . . . . weeks after the Contract Date are . . . . . . . . . . . . . . . . . . . . (either an amount or a percentage of the payment otherwise due)

- An advanced payment bond <u>is/is not</u> required

If Option K is used

- Payment for the items or activities listed below will be made in the currencies stated

| Items or activities | Currency | Maximum payment |
|---|---|---|
| . . . . . . . . . . . . . . | . . . . . . . . . . | . . . . . . . . . . . . . . . . . . . . . . . . . . . . . . . . . . . . . . |
| . . . . . . . . . . . . . . | . . . . . . . . . . | . . . . . . . . . . . . . . . . . . . . . . . . . . . . . . . . . . . . . . |
| . . . . . . . . . . . . . . | . . . . . . . . . . | . . . . . . . . . . . . . . . . . . . . . . . . . . . . . . . . . . . . . . |

- The *exchange rates* are those published in . . . . . . . . . . . . . . . . . . . . . . . . . . . . . . . .

  on . . . . . . . . . . . . . . . . . . . . . . . . . . . (date)

If Option L is used

- The *subcontract completion date* for each *section* of the *subcontract works* is

| Section | Description | Subcontract completion date |
|---|---|---|
| 1 | . . . . . . . . . . . . . . . . . . | . . . . . . . . . . . . . . . . . . . . . . . . . . . . . . . . . . . . |
| 2 | . . . . . . . . . . . . . . . . . . | . . . . . . . . . . . . . . . . . . . . . . . . . . . . . . . . . . . . |
| 3 | . . . . . . . . . . . . . . . . . . | . . . . . . . . . . . . . . . . . . . . . . . . . . . . . . . . . . . . |
| 4 | . . . . . . . . . . . . . . . . . . | . . . . . . . . . . . . . . . . . . . . . . . . . . . . . . . . . . . . |
| 5 | . . . . . . . . . . . . . . . . . . | . . . . . . . . . . . . . . . . . . . . . . . . . . . . . . . . . . . . |

If Options L and Q are used together

- The bonuses for the *section*s of the *subcontract works* are

| Section | Description | Amount per day |
|---|---|---|
| 1 | . . . . . . . . . . . . . . . . . . | . . . . . . . . . . . . . . . . . . . . . . . . . . . . . . . . . . . . |
| 2 | . . . . . . . . . . . . . . . . . . | . . . . . . . . . . . . . . . . . . . . . . . . . . . . . . . . . . . . |
| 3 | . . . . . . . . . . . . . . . . . . | . . . . . . . . . . . . . . . . . . . . . . . . . . . . . . . . . . . . |
| 4 | . . . . . . . . . . . . . . . . . . | . . . . . . . . . . . . . . . . . . . . . . . . . . . . . . . . . . . . |
| 5 | . . . . . . . . . . . . . . . . . . | . . . . . . . . . . . . . . . . . . . . . . . . . . . . . . . . . . . . |

If Options L and R are used together

- Delay damages for the *section*s of the *subcontract works* are

| Section | Description | Amount per day |
|---|---|---|
| 1 | . . . . . . . . . . . . . . . . . . | . . . . . . . . . . . . . . . . . . . . . . . . . . . . . . . . . . . . |
| 2 | . . . . . . . . . . . . . . . . . . | . . . . . . . . . . . . . . . . . . . . . . . . . . . . . . . . . . . . |
| 3 | . . . . . . . . . . . . . . . . . . | . . . . . . . . . . . . . . . . . . . . . . . . . . . . . . . . . . . . |
| 4 | . . . . . . . . . . . . . . . . . . | . . . . . . . . . . . . . . . . . . . . . . . . . . . . . . . . . . . . |
| 5 | . . . . . . . . . . . . . . . . . . | . . . . . . . . . . . . . . . . . . . . . . . . . . . . . . . . . . . . |

If Option N is used

- The proportions used to calculate the Price Adjustment Factor are

  0 · . . . . . . . . . . . .      linked to the index for . . . . . . . . . . . . . . . . . . . . .

  0 · . . . . . . . . . . . .                                    . . . . . . . . . . . . . . . . . . . .

  0 · . . . . . . . . . . . .                                    . . . . . . . . . . . . . . . . . . . .

  0 · . . . . . . . . . . . .                                    . . . . . . . . . . . . . . . . . . . .

  0 · . . . . . . . . . . . .                                    . . . . . . . . . . . . . . . . . . . .

  0 · . . . . . . . . . . . .                                    . . . . . . . . . . . . . . . . . . . .

  0 · . . . . . . . . . . . .      non-adjustable

  _____

  1 · 00

- The *base date* for indices is . . . . . . . . . . . . . . . . . . . . . . . . . . . . . . . . . . . . . . . . . . . . .

- The indices are those prepared by . . . . . . . . . . . . . . . . . . . . . . . . . . . . . . . . . . . . . . .

If Option P is used

- The *retention free amount* is . . . . . . . . . . . . . . . . . . . . . . . . . . . . . . . . . . . . . . . . . . .

- The *retention percentage* is . . . . . . . . . . %

If Option Q is used

- The bonus for the whole of the *subcontract works* is . . . . . . . . . . . . . . . . . per day

If Option R is used (whether or not Option L is also used)

- Delay damages for the whole of the *subcontract works* are . . . . . . . . . . . . . per day

If Option S is used

- The amounts for low performance damages are

  | Amount | Performance level |
  |---|---|
  | . . . . . . . . . . . . . . . . . . . . . . . . . | for . . . . . . . . . . . . . . . . . . . . . . . . . . . |
  | . . . . . . . . . . . . . . . . . . . . . . . . . | for . . . . . . . . . . . . . . . . . . . . . . . . . . . |
  | . . . . . . . . . . . . . . . . . . . . . . . . . | for . . . . . . . . . . . . . . . . . . . . . . . . . . . |
  | . . . . . . . . . . . . . . . . . . . . . . . . . | for . . . . . . . . . . . . . . . . . . . . . . . . . . . |

If Option U is used

- These special conditions apply

  . . . . . . . . . . . . . . . . . . . . . . . . . . . . . . . . . . . . . . . . . . . . . . . . . . . . . . . . . . . . . . . . . . . .

  . . . . . . . . . . . . . . . . . . . . . . . . . . . . . . . . . . . . . . . . . . . . . . . . . . . . . . . . . . . . . . . . . . . .

# Part two - Data provided by the *Subcontractor*

**Statements given in all subcontracts**

- The *Subcontractor* is

    Name . . . . . . . . . . . . . . . . . . . . . . . . . . . . . . . . . . . . . . . . . . . . . . . . . . . . . . . . . . .

    Address . . . . . . . . . . . . . . . . . . . . . . . . . . . . . . . . . . . . . . . . . . . . . . . . . . . . . . . . . .

    . . . . . . . . . . . . . . . . . . . . . . . . . . . . . . . . . . . . . . . . . . . . . . . . . . . . . . . . . . . . . . . . .

- The *fee percentage* is. . . . . . . . . . . . . . . . . . . . . . . . . . . . . . . . . . . . . . . . . . . . . .

- The *working areas* are the Site and . . . . . . . . . . . . . . . . . . . . . . . . . . . . . . . . . . .

- The key people are

    (1) Name . . . . . . . . . . . . . . . . . . . . . . . . . . . . . . . . . . . . . . . . . . . . . . . . . . . . . . . . .

    Job . . . . . . . . . . . . . . . . . . . . . . . . . . . . . . . . . . . . . . . . . . . . . . . . . . . . . . . . . . . . . .

    Responsibilities . . . . . . . . . . . . . . . . . . . . . . . . . . . . . . . . . . . . . . . . . . . . . . . . . . . .

    . . . . . . . . . . . . . . . . . . . . . . . . . . . . . . . . . . . . . . . . . . . . . . . . . . . . . . . . . . . . . . . . .

    Qualifications . . . . . . . . . . . . . . . . . . . . . . . . . . . . . . . . . . . . . . . . . . . . . . . . . . . . . .

    Experience . . . . . . . . . . . . . . . . . . . . . . . . . . . . . . . . . . . . . . . . . . . . . . . . . . . . . . . .

    . . . . . . . . . . . . . . . . . . . . . . . . . . . . . . . . . . . . . . . . . . . . . . . . . . . . . . . . . . . . . . . . .

    (2) Name . . . . . . . . . . . . . . . . . . . . . . . . . . . . . . . . . . . . . . . . . . . . . . . . . . . . . . . . .

    Job . . . . . . . . . . . . . . . . . . . . . . . . . . . . . . . . . . . . . . . . . . . . . . . . . . . . . . . . . . . . . .

    Responsibilities . . . . . . . . . . . . . . . . . . . . . . . . . . . . . . . . . . . . . . . . . . . . . . . . . . . .

    . . . . . . . . . . . . . . . . . . . . . . . . . . . . . . . . . . . . . . . . . . . . . . . . . . . . . . . . . . . . . . . . .

    Qualifications . . . . . . . . . . . . . . . . . . . . . . . . . . . . . . . . . . . . . . . . . . . . . . . . . . . . . .

    Experience . . . . . . . . . . . . . . . . . . . . . . . . . . . . . . . . . . . . . . . . . . . . . . . . . . . . . . . .

    . . . . . . . . . . . . . . . . . . . . . . . . . . . . . . . . . . . . . . . . . . . . . . . . . . . . . . . . . . . . . . . . .

**Optional statements**

If the *Subcontractor* is to provide Subcontract Works Information for his design

- The Subcontract Works Information for the *Subcontractor*'s design is in

    . . . . . . . . . . . . . . . . . . . . . . . . . . . . . . . . . . . . . . . . . . . . . . . . . . . . . . . . . . . . . . . . .

    . . . . . . . . . . . . . . . . . . . . . . . . . . . . . . . . . . . . . . . . . . . . . . . . . . . . . . . . . . . . . . . . .

    . . . . . . . . . . . . . . . . . . . . . . . . . . . . . . . . . . . . . . . . . . . . . . . . . . . . . . . . . . . . . . . . .

    . . . . . . . . . . . . . . . . . . . . . . . . . . . . . . . . . . . . . . . . . . . . . . . . . . . . . . . . . . . . . . . . .

    . . . . . . . . . . . . . . . . . . . . . . . . . . . . . . . . . . . . . . . . . . . . . . . . . . . . . . . . . . . . . . . . .

    . . . . . . . . . . . . . . . . . . . . . . . . . . . . . . . . . . . . . . . . . . . . . . . . . . . . . . . . . . . . . . . . .

If the *Subcontractor* is to provide a programme

- The programme numbered . . . . . . . . . . . . . . . . . . . . . is the Accepted Programme

If the *Subcontractor* is to provide the *subcontract completion date* for the whole of the *subcontract works*

• The *subcontract completion date* for the whole of the *subcontract works* is . . . . . . .

If Option A or C is used

• The *activity schedule* is . . . . . . . . . . . . . . . . . . . . . . . . . . . . . . . . . . . . . . . . . . . . . . . .

If Option B or D is used

• The *bill of quantities* is . . . . . . . . . . . . . . . . . . . . . . . . . . . . . . . . . . . . . . . . . . . . . . . . .

If Option A, B, C or D is used

• The tendered total of the Prices is . . . . . . . . . . . . . . . . . . . . . . . . . . . . . . . . . . . . . . .

**Cost component data**

• The percentage for Equipment depreciation and maintenance is . . . . . . . . . . . . %
(not used for the short method)

• The percentage for Working Areas overheads is . . . . . . . . . . . . . . . . . . . . . . . . %
(not used for the short method)

• The hourly rates for Actual Cost of manufacture and fabrication outside the Working Areas are

• Category of employee          Hourly rate

. . . . . . . . . . . . . . . . . . . . . . . . . . . .          . . . . . . . . . . . . . . . . . . . . . . . . . . . . . . . . . . .

. . . . . . . . . . . . . . . . . . . . . . . . . . . .          . . . . . . . . . . . . . . . . . . . . . . . . . . . . . . . . . . .

. . . . . . . . . . . . . . . . . . . . . . . . . . . .          . . . . . . . . . . . . . . . . . . . . . . . . . . . . . . . . . . .

. . . . . . . . . . . . . . . . . . . . . . . . . . . .          . . . . . . . . . . . . . . . . . . . . . . . . . . . . . . . . . . .

The percentage for manufacture and fabrication overheads is. . . . . . . . . . . . . . . . %

• The hourly rates for Actual Cost of design outside the Working Areas are

• Category of employee          Hourly rate

. . . . . . . . . . . . . . . . . . . . . . . . . . . .          . . . . . . . . . . . . . . . . . . . . . . . . . . . . . . . . . . . .

. . . . . . . . . . . . . . . . . . . . . . . . . . . .          . . . . . . . . . . . . . . . . . . . . . . . . . . . . . . . . . . . .

. . . . . . . . . . . . . . . . . . . . . . . . . . . .          . . . . . . . . . . . . . . . . . . . . . . . . . . . . . . . . . . . .

. . . . . . . . . . . . . . . . . . . . . . . . . . . .          . . . . . . . . . . . . . . . . . . . . . . . . . . . . . . . . . . . .

• The percentage for design overheads is. . . . . . . . . . . . . . . . . . . . . . . . . . . . . . . . %

• The categories of employees whose travelling expenses to and from the Working Areas are included in Actual Cost are

. . . . . . . . . . . . . . . . . . . . . . . . . . . . . . . . . . . . . . . . . . . . . . . . . . . . . . . . . . . . . . . . . . . .

. . . . . . . . . . . . . . . . . . . . . . . . . . . . . . . . . . . . . . . . . . . . . . . . . . . . . . . . . . . . . . . . . . . .

. . . . . . . . . . . . . . . . . . . . . . . . . . . . . . . . . . . . . . . . . . . . . . . . . . . . . . . . . . . . . . . . . . . .

. . . . . . . . . . . . . . . . . . . . . . . . . . . . . . . . . . . . . . . . . . . . . . . . . . . . . . . . . . . . . . . . . . . .

**Cost component data for the short method**

- The percentage for people overheads is . . . . . . . . . . . . . . . . . . . . . . . . . . . . . . . . %

- The published list of Equipment is the last edition of the list published by

  . . . . . . . . . . . . . . . . . . . . . . . . . . . . . . . . . . . . . . . . . . . . . . . . . . . . . . . . . . . . .

- The percentage for adjustment for listed Equipment is . . . . . . . . . . . . . . . . . . . %

- The list of Equipment and rates is

| Description | Size or capacity | Rate |
|---|---|---|
| . . . . . . . . . . . . . . . . . . . | . . . . . . . . . . . . . . . . . . . . . . | . . . . . . . . . . . . . . . . . |
| . . . . . . . . . . . . . . . . . . . | . . . . . . . . . . . . . . . . . . . . . . | . . . . . . . . . . . . . . . . . |
| . . . . . . . . . . . . . . . . . . . | . . . . . . . . . . . . . . . . . . . . . . | . . . . . . . . . . . . . . . . . |
| . . . . . . . . . . . . . . . . . . . | . . . . . . . . . . . . . . . . . . . . . . | . . . . . . . . . . . . . . . . . |
| . . . . . . . . . . . . . . . . . . . | . . . . . . . . . . . . . . . . . . . . . . | . . . . . . . . . . . . . . . . . |
| . . . . . . . . . . . . . . . . . . . | . . . . . . . . . . . . . . . . . . . . . . | . . . . . . . . . . . . . . . . . |